Holt Mathematics

Chapter 2 Resource Book

HOLT, RINEHART AND WINSTON
A Harcourt Education Company
Orlando • Austin • New York • San Diego • London

Printed in the United States of America

ISBN 0-03-078297-X

1 2 3 4 5 170 09 08 07 06

CONTENTS

Blackline Masters

Date ____________

Dear Family,

In this chapter, your child will work with integers and solve one-step equations containing integers. Your child will also learn how to find the prime factorization of a number, and how to find the greatest common factor and the least common multiple of two or more numbers. Your child will then work with rational numbers.

Integers are the set of whole numbers and their opposites. For instance, the opposite of 8 is –8.

Understanding how to work with integers is important for solving algebraic equations. You can use a number line to understand how to **add and subtract integers.**

Add 3 + (−5).

Start at 0. Move right 3 units to add 3.

Move left 5 units to add −5.

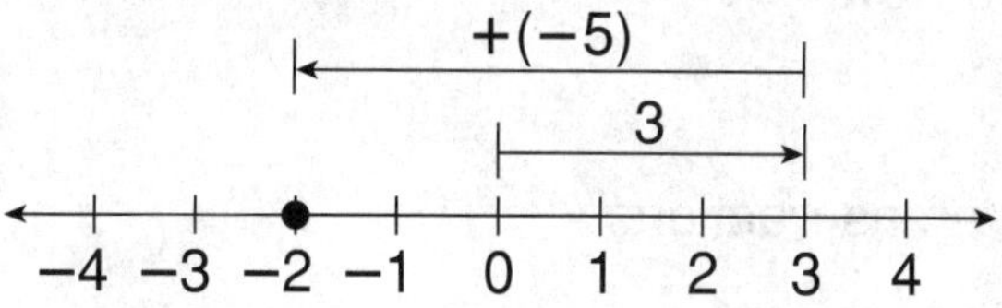

$3 + (-5) = -2$

Subtract −1 − (−3).

To subtract an integer, add its opposite.

$-1 - (-3) = -1 + 3$

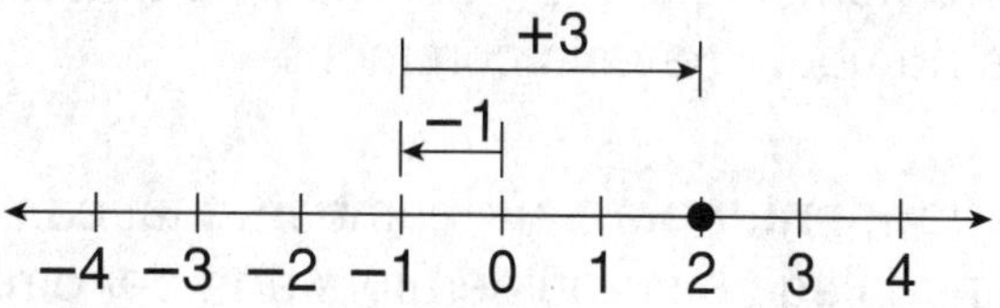

$-1 - (-3) = 2$

To **multiply or divide integers,** follow these rules.

When both integers have the same sign, the answer is positive.

$2 \times 8 = 16$ $\qquad$ $-16 \div (-8) = 2$

When the two integers have different signs, the answer is negative.

$-2 \times 8 = -16$ $\qquad$ $16 \div (-8) = -2.$

You can **solve equations with integers** by adding, subtracting, multiplying, or dividing.

Solve −3 + *y* = −5.

$$-3 + y = -5$$
$$\underline{+3 \qquad\quad +3}$$
$$y = -2$$

← Add 3 to both sides to isolate the variable *y*.

Solve −120 = 6*x*.

$$-120 = 6x$$
$$\frac{-120}{6} = \frac{6x}{6}$$

← Divide each side by 6 to isolate the variable *x*.

Factors are numbers that are multiplied to get a **product.**

To write the **prime factorization** of a number, write the number as the product of prime numbers. A **prime number** has only two factors, itself and 1. One way to find the prime factorization of a number is to use a factor tree like the one at the right.

24	Write 24 as the product
$4 \cdot 6$	of two factors.
$2 \cdot 2 \cdot 2 \cdot 3$	Keep factoring until all the factors are prime.

The prime factorization of 24 is $2 \cdot 2 \cdot 2 \cdot 3$, or $2^3 \cdot 3$.

The **greatest common factor (GCF)** of two or more numbers is the greatest number that divides evenly into each number. One way to find the GCF of two or more numbers is to list all the factors of each number.

Find the GCF of 20 and 32.
Factors of 20: 1, 2, 4, 5, 10, 20
Factors of 32: 1, 2, 4, 8, 16, 32
The GCF is 4.

A **multiple** of a number is the product of that number and a whole number. The **least common multiple (LCM)** of two or more numbers is the least multiple that the numbers have in common.

Find the LCM of 6 and 8.
Multiples of 6: 6, 12, 18, 24, ...
Multiples of 8: 8, 16, 24, 32, ...
The LCM is 24.

Rational numbers are numbers that can be written as fractions using integers for numerators and denominators.

$15 = \frac{15}{1}$ $\quad -5 = \frac{-5}{1}$ $\quad 0.75 = \frac{3}{4}$ $\quad -1.25 = \frac{-5}{4}$

Equivalent fractions are the same number expressed in different ways.

$\frac{1}{2} = \frac{2}{4} = \frac{3}{6}$

To **convert** a fraction to a decimal, divide the numerator by the denominator.

$\frac{3}{4}$

$$\begin{array}{r} 0.75 \\ 4\overline{)3.00} \\ -\,28 \\ \hline 20 \\ -\,20 \\ \hline 0 \end{array}$$

$\frac{3}{4} = 0.75$

For additonal resources, visit go.hrw.com and enter the keyword MS7 Parent.

Name ______________________ Date ____________ Class ____________

LESSON 2-1

Practice A

Integers

Graph each integer and its opposite on a number line.

1. 3

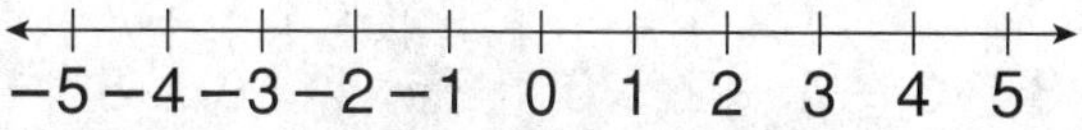

2. –5

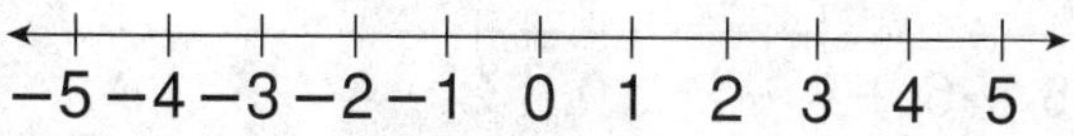

Use the number line to compare the integers. Write < or >.

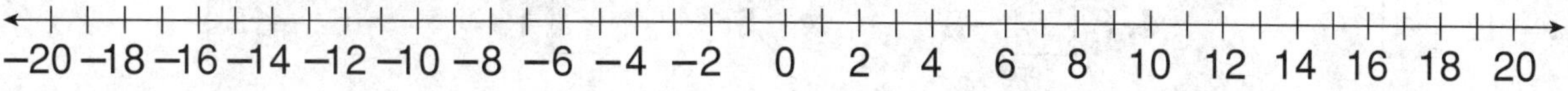

3. –8 ☐ 7 **4.** 4 ☐ –7 **5.** –6 ☐ –16 **6.** –11 ☐ 11

Graph the integers on a number line. Then write them in order from least to greatest.

7. –6; 3; –5; 8

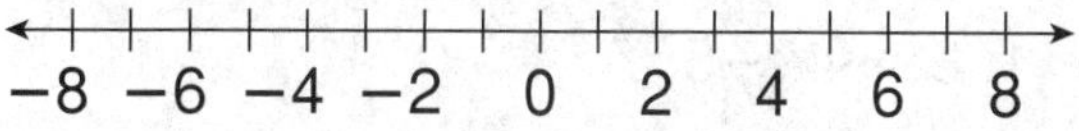

____

8. 6; –7; –8; 0

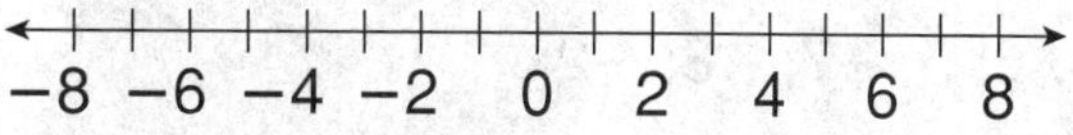

____

Use a number line to find each absolute value.

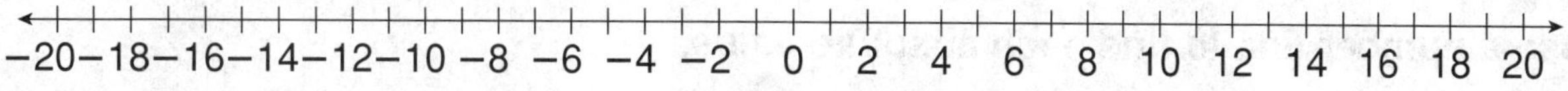

9. $|-6|$ ______ **10.** $|2|$ ______ **11.** $|-1|$ ______ **12.** $|8|$ ______

13. $|-9|$ ______ **14.** $|3|$ ______ **15.** $|-4|$ ______ **16.** $|10|$ ______

17. $|-15|$ ______ **18.** $|20|$ ______ **19.** $|-13|$ ______ **20.** $|17|$ ______

21. The windchill on a cold day made it feel like 5 degrees below zero outside. Write this temperature as an integer.

____

22. A baby gained 15 pounds from birth to his first birthday. Write this amount as an integer.

____

Name ______________________ Date ____________ Class ____________

LESSON **2-1**

Practice B

Integers

Graph each integer and its opposite on a number line.

1. 8

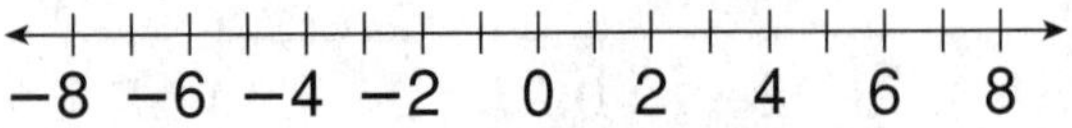

2. −7

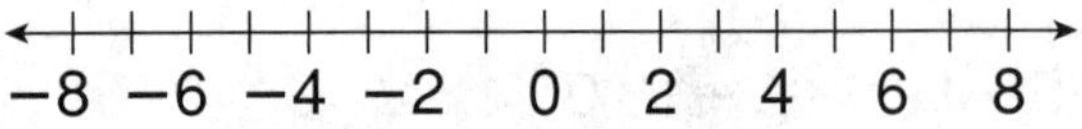

Compare the integers. Use $<$ or $>$.

3. −15 ☐ −7 **4.** 8 ☐ −8 **5.** −14 ☐ 13 **6.** −18 ☐ −20

Use a number line to order the integers from least to greatest.

7. −1; 4; −5; 7; −3

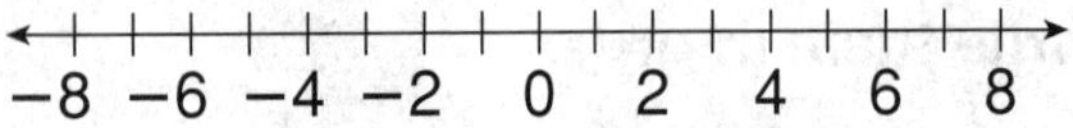

____

8. −6; 8; 0; 4; −2

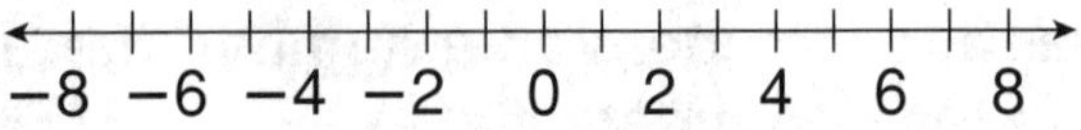

____

9. 6; 5; −7; −8; −2

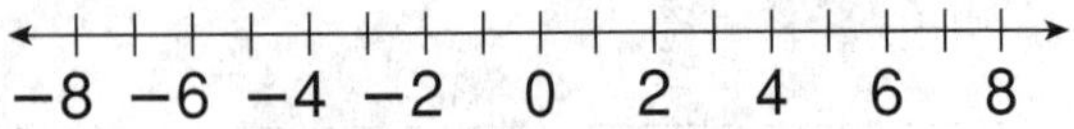

____

10. 1; 3; −4; −5; 7

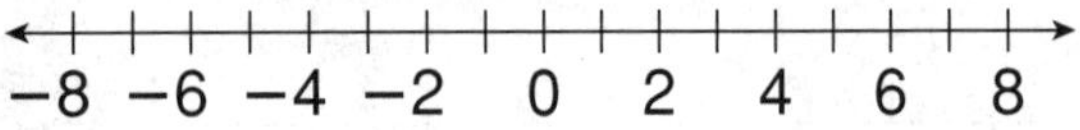

____

Use a number line to find each absolute value.

11. $|-18|$ ______ **12.** $|11|$ ______ **13.** $|-25|$ ______ **14.** $|19|$ ______

15. $|-10|$ ______ **16.** $|16|$ ______ **17.** $|22|$ ______ **18.** $|-14|$ ______

19. $|9|$ ______ **20.** $|-24|$ ______ **21.** $|-7|$ ______ **22.** $|17|$ ______

23. Christy dove to a depth of 12 feet below the surface of the water. Write the depth as an integer.

____

24. The highest point in North Carolina is Mt. Mitchell, with a height of 6,684 feet. Write the height of Mt. Mitchell as an integer.

____

Name ______________________________ Date ____________ Class ____________

LESSON **2-1**

Practice C

Integers

Graph each integer and its opposite on a number line.

1. 15

2. −4

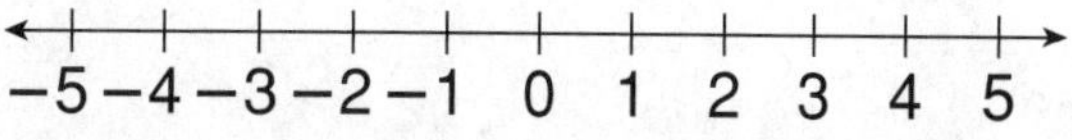

Compare the integers. Use < or >.

3. −15 ☐ 15 **4.** −7 ☐ −9 **5.** 12 ☐ −17 **6.** −13 ☐ 9

Use a number line to order the integers from least to greatest.

7. −2; 3; −6; −1; 4

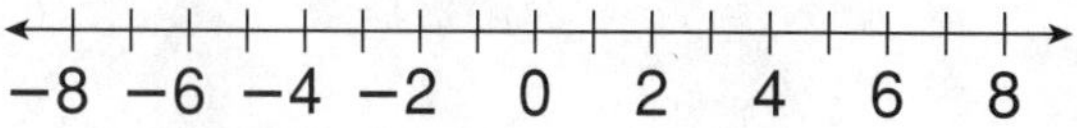

____

8. 7; −5; 6; 0; −7

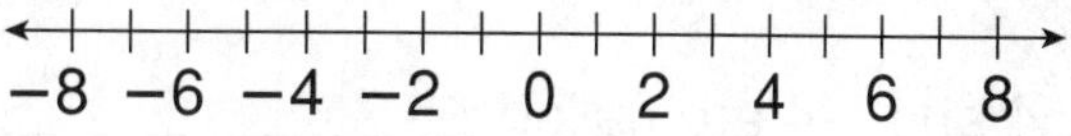

____

Find each absolute value.

9. $|-45|$ ______ **10.** $|38|$ ______ **11.** $|-99|$ ______ **12.** $|22|$ ______

13. $|-19|$ ______ **14.** $|375|$ ______ **15.** $|-59|$ ______ **16.** $|-84|$ ______

Compare. Write <, >, or =.

17. $|-50|$ ☐ 50 **18.** 28 ☐ $|-29|$ **19.** −43 ☐ $|-37|$ **20.** $|48|$ ☐ 53

21. $|-79|$ ☐ $|65|$ **22.** −92 ☐ $|-92|$ **23.** $|-83|$ ☐ 82 **24.** $|34|$ ☐ $|-62|$

25. A company reported a net gain of $750,000 for the first quarter and a net loss of $2,000,000 for the second quarter. Write each amount as an integer.

____

26. The record low temperature in Virginia through 1997 was 30°F below 0. The record high temperature was 110°F. Write each temperature as an integer.

____

Name ______________________ Date ____________ Class ____________

LESSON **2-1**

Reteach

Integers

This number line shows integers.

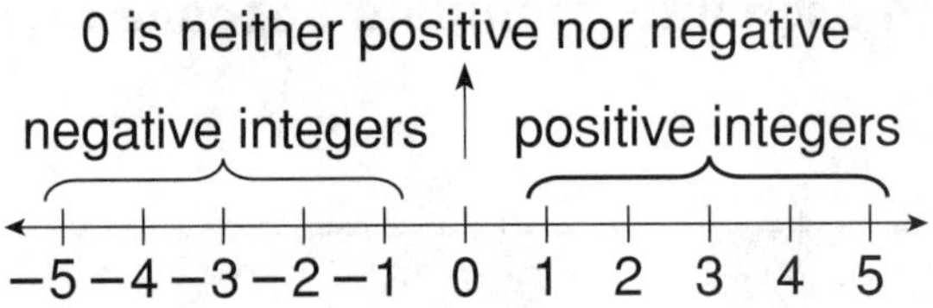

Every integer has an **opposite** integer. A number and its opposite are the same distance from 0.

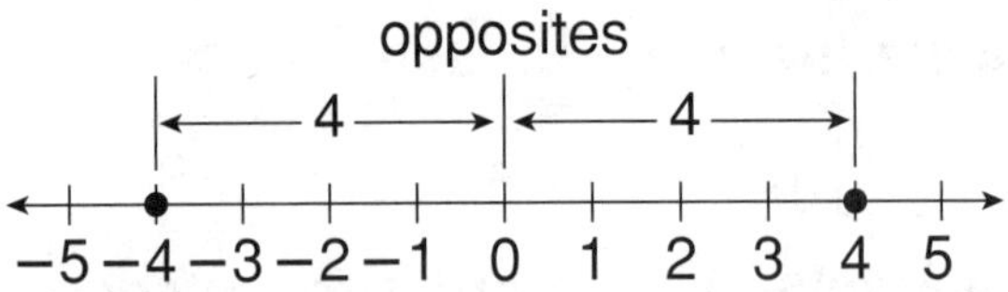

1. How many units is 4 from 0? ______

2. How many units is −4 from 0? ______

3. 4 and −4 are called ______________________________.

Graph each integer and its opposite on a number line.

4. 2

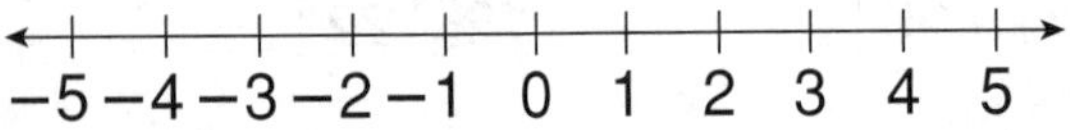

5. −3

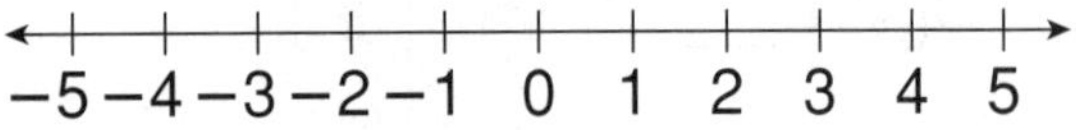

You can use a number line to compare and order numbers. The numbers get greater as you move to the right on the number line.

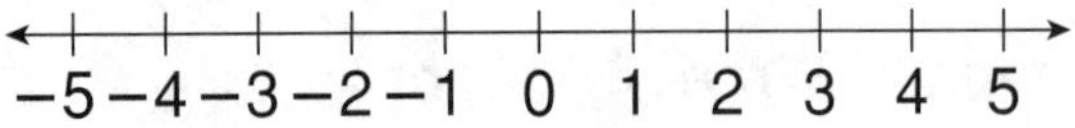

6. What is the order from least to greatest of −1, 2, and −3? ______________

Write the integers in order from least to greatest.

7. −2; −6; 4

____

8. −3; 7; 1

____

The absolute value of an integer is its distance from 0 on a number line. −5 is 5 units from 0. The absolute value of −5 is 5. You write $|-5| = 5$.

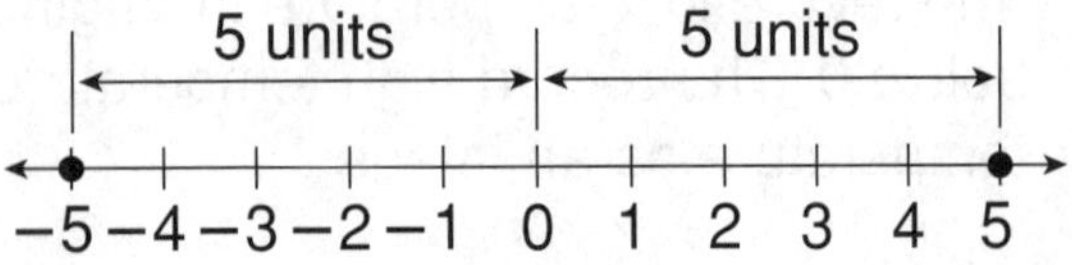

9. How many units from 0 is −3? ______________________

Name ______________________ Date ____________ Class ____________

LESSON **2-1**

Challenge

Order That Integer!

Use two < signs to correctly order each set of integers.

1. −12; 12; −11

____

2. −12; −210; −201

____

3. |−3|; |−6|; |1|

____

4. |−38|; |3|; −15

____

Use two > signs to correctly order each set of integers.

5. −4; −40; 4

____

6. 0; −8; |−18|

____

7. |−75|; |12|; 15

____

8. −16; −12; 1

____

Find each missing integer. Write its corresponding letter in each blank to answer the riddle.

A	B	C	E	V	T	L	S	U	O	N
−4	−6	7	6	−3	0	−7	−1	1	2	−2

WHAT DID THE CAR DEALER SAY TO THE CUSTOMER?

THIS DEAL IS

___	___
−5; ___; −3	−3; ___; −1

___	___	___	___	___	___	___	___
___; −3; −2	−7; ___; −5	___; 0; 1	0; 1; ___	−9; −8; ___	−1; 0; ___	−1; ___; 1	5; ___; 7

___	___	___	___	___
___; −2; −1	−6; −5; ___	___; −6; −5	0; ___; 2	___; 7; 8

Name ______________________________ Date __________ Class __________

LESSON 2-1

Problem Solving

Integers

Write the correct answer.

1. The coldest place on record in the United States was in Alaska in 1971. It was 80°F below zero. Write this temperature as an integer.

____

2. The temperature outside was −4°F at Jared's house and −8°F at Mario's house. Where was the temperature warmer?

____

3. A small business reported a net loss of $62,500 during its first year. In its second year, it reported a profit of $34,100. Write each amount as an integer.

____

4. For one day, Lacy recorded the low temperatures in five U.S. cities. The temperatures were 5°C, −1°C, −3°C, 2°C, and 0°C. Write the temperatures in order from least to greatest.

____

Choose the letter for the best answer.

5. Which number is not an integer?

$-3; 5; \frac{1}{5}; 0$

A −3 **C** $\frac{1}{5}$

B 5 **D** 0

6. Basha says |5| and |−5| are the same number. Danny says |5| and |−5| are different numbers. Kim says |5| and |−5| both equal 0. Who is correct?

F Basha

G Danny

H Kim

J They are all wrong.

7. Use the table at right. Which continent has the highest point?

A Asia **C** Africa

B South America **D** Australia

8. Use the table at right. Which continent has the lowest point?

F Europe **H** North America

G Australia **J** Asia

Continent	Highest Point (ft)	Lowest Point (ft)
North America	20,320	−282
South America	22,834	−131
Africa	19,340	−512
Asia	29,028	−1,339
Australia	7,310	−52
Europe	18,510	−92

Name ______________________________ Date ____________ Class ____________

LESSON 2-1

Reading Strategies

Use a Graphic Organizer

Integers

Definition	**Facts**
The set of whole numbers and their opposites	• Each number can be paired with its opposite. The opposite of 2 is −2. The opposite of −3 is 3. • Zero is its own opposite.
Examples	**Non-examples**
0, 2, 5, 9, 13, −3, −7, −12, −17	$\frac{2}{3}$, $\frac{11}{5}$, $2\frac{5}{8}$, 0.5, 0.23, 1.05, 3.61

Answer each question.

1. What are integers?

____

2. Write the opposite of 6. ____________

3. Write the opposite of 10. ____________

4. Write the opposite of 0. ____________

5. Write the opposite of −8. ____________

6. Write the opposite of −3. ____________

Write "integer" or "not an integer" for the following numbers.

7. −9 ____________________

8. $\frac{5}{7}$ ____________________

9. 0.1 ____________________

10. 42 ____________________

Name ______________________________ Date ____________ Class ____________

LESSON 2-1

Puzzles, Twisters & Teasers

Food for Thought!

Circle words from the list in the word search. Then find a word in the word search that answers the riddle. Write the word on the lines below.

negative	opposite	number	integer	absolute
value	set	express	whole	zero

M	H	T	E	F	V	C	D	E	S	W	I	C	F	R
O	P	P	O	S	I	T	E	V	F	R	N	B	H	U
X	Z	O	I	E	S	W	A	X	N	M	T	A	W	H
N	E	G	A	T	I	V	E	C	P	U	E	I	H	Q
U	R	A	S	D	F	G	H	V	L	R	G	P	O	K
M	O	Z	X	C	V	B	N	A	R	J	E	L	L	Y
B	P	O	I	U	W	H	O	L	E	K	R	S	E	M
E	D	F	G	H	J	K	L	U	W	E	R	T	S	X
R	A	B	S	O	L	U	T	E	Z	A	W	Q	E	C

What kind of fish goes best with peanut butter?

___ ___ ___ ___ ___

Name ______________________ Date ___________ Class ___________

LESSON **2-2**
Practice A
Adding Integers

Show the addition on the number line. Then write the sum.

1. 2 + (−3)

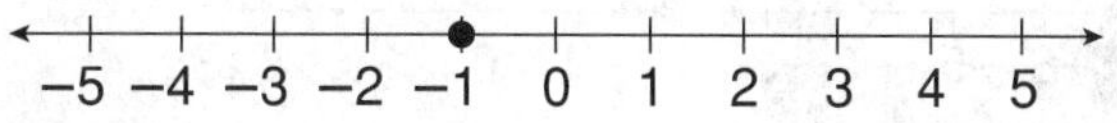

____

2. −3 + (−4)

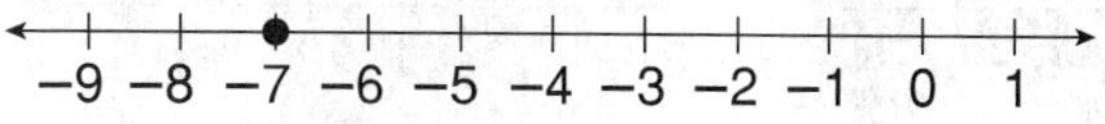

____

Find each sum.

3. −4 + (−9) ________

4. 7 + (−8) ________

5. −2 + 1 ________

6. 6 + (−9) ________

7. 5 + 7 ________

8. 9 + (−5) ________

9. (−1) + 9 ________

10. −9 + (−7) ________

11. 2 + (−7) ________

12. −6 + (−4) ________

13. 3 + 2 ________

14. −2 + 6 ________

Evaluate $e + f$ for the given values.

15. $e = 9$, $f = -2$ ________

16. $e = -4$, $f = -6$ ________

17. $e = 6$, $f = -1$ ________

18. $e = -3$, $f = 2$ ________

19. $e = 8$, $f = -6$ ________

20. $e = -2$, $f = -3$ ________

21. The temperature dropped 13°F in 7 hours. The final temperature was −2°F. What was the starting temperature?

____

22. A football team gains 8 yards in one play, then loses 5 yards in the next. What is the team's total yardage?

____

Name ______________________ Date __________ Class __________

LESSON 2-2

Practice B

Adding Integers

Use a number line to find each sum.

1. $-1 + 5$

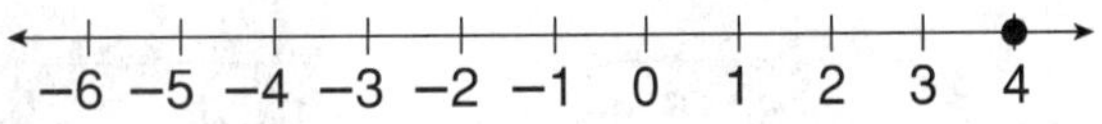

____

2. $4 + (-6)$

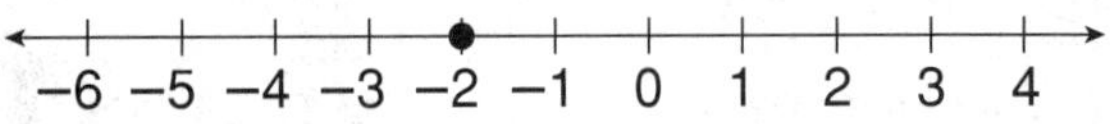

____

Find each sum.

3. $-51 + (-9)$ ________

4. $27 + (-6)$ ________

5. $1 + (-30)$ ________

6. $15 + (-25)$ ________

7. $50 + (-7)$ ________

8. $-19 + (-15)$ ________

9. $(-23) + 9$ ________

10. $-19 + (-21)$ ________

11. $-17 + 11$ ________

12. $20 + (-8)$ ________

13. $(-15) + (-7)$ ________

14. $12 + (-14)$ ________

Evaluate $e + f$ for the given values.

15. $e = 9$, $f = -24$ ________

16. $e = -17$, $f = -7$ ________

17. $e = 32$, $f = -19$ ________

18. $e = -15$, $f = -15$ ________

19. $e = -20$, $f = 20$ ________

20. $e = -30$, $f = 12$ ________

21. The temperature rose 9°F in 3 hours. If the starting temperature was −5°F, what was the final temperature?

____

22. Matt is playing a game. He gains 7 points, loses 10 points, gains 2 points, and then loses 8 points. What is his final score?

____

Name ____________________ Date __________ Class __________

LESSON 2-2

Practice C
Adding Integers

Use a number line to find each sum.

1. $-7 + 10$

____

2. $-2 + (-6)$

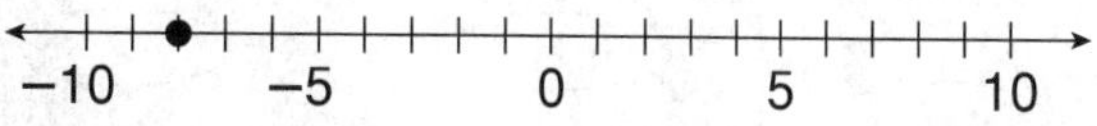

____

Find each sum.

3. $-45 + 9$ __________

4. $37 + (-60)$ __________

5. $-11 + 40$ __________

6. $95 + (-25)$ __________

7. $56 + (-37)$ __________

8. $-79 + (-15)$ __________

9. $(-63) + 49$ __________

10. $-49 + (-47)$ __________

Evaluate each expression for $a = -15$, $b = -8$, and $c = 6$.

11. $14 + b$ ______

12. $c + (-8)$ ______

13. $a + b$ ______

14. $c + 5 + a$ ______

Evaluate $e + f$ for the given values.

15. $e = 90$, $f = -52$ __________

16. $e = -57$, $f = -23$ __________

17. $e = -61$, $f = 32$ __________

Compare. Write $<$, $>$, or $=$.

18. $25 + (-25)$ ☐ 0

19. $32 + (-19)$ ☐ -51

20. $24 + (-42)$ ☐ 66

21. The temperature dropped 10°F in 5 hours. If the starting temperature was 9°F, what was the final temperature?

____

22. The closing price of a stock gained 2 points on Monday, lost 5 points on Tuesday, lost 1 point on Wednesday, gained 4 points on Thursday, and lost 6 points on Friday. What was the net gain or loss for the week?

____

Name ______________________ Date ____________ Class ____________

LESSON 2-2

Reteach

Adding Integers

This balance scale "weighs" positive and negative numbers. Negative numbers go on the left of the balance, and positive numbers go on the right.

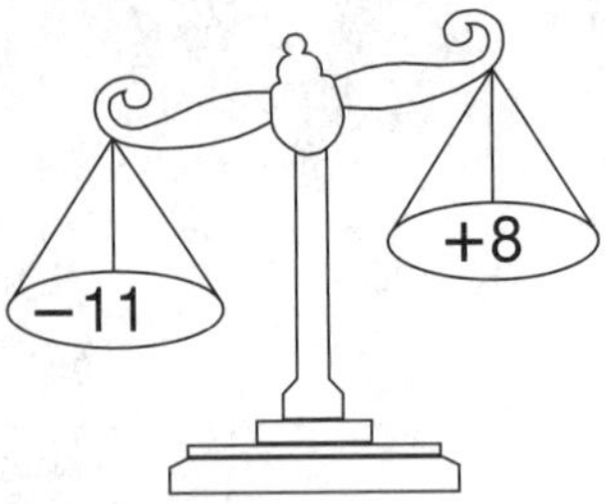

Find −11 + 8.

The scale will tip to the left side because the sum of −11 and +8 is negative.

$-11 + 8 = -3$

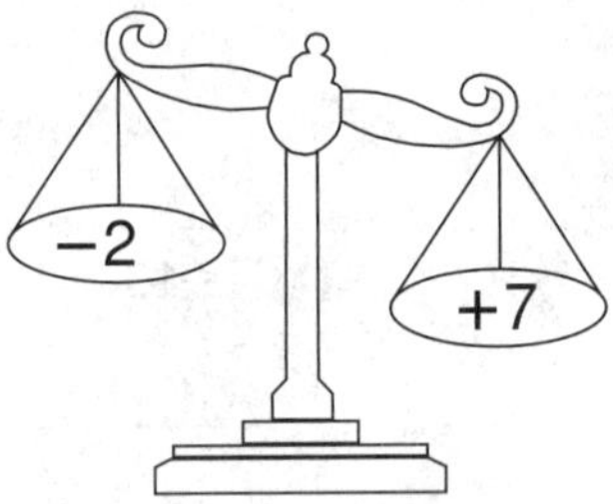

Find −2 + 7.

The scale will tip to the right side because the sum of −2 and +7 is positive.

$-2 + 7 = 5$

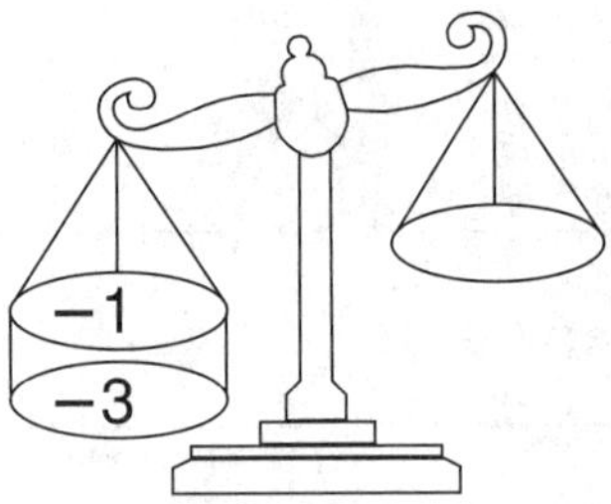

Find −1 + (−3).

Both −1 and −3 go on the left side. The scale will tip to the left side because the sum of −1 and −3 is negative.

$-1 + (-3) = -4$

Find 3 + (−9).

1. Should you add or subtract? ______________________

2. Will the sum be positive or negative? ______________________

$3 + (-9) = -6$

the sign of the integer with the greatest absolute value

$|9| - |3|$

Find −5 + (−8).

3. Should you add or subtract? ______________________

4. Will the sum be positive or negative? ______________________

5. $-5 + (-8) =$ ______________________

Add.

6. $7 + -3 =$ ______ **7.** $-2 + -3 =$ ______ **8.** $-5 + 4 =$ ______

9. $-3 + -1 =$ ______ **10.** $-7 + 9 =$ ______ **11.** $4 + -9 =$ ______

Name ______________________ Date __________ Class __________

LESSON 2-2

Challenge

Alphabet Addition

Find the value of each word. Each vowel has a value as shown in the table. All consonants have a value of 2.

A	E	I	O	U
−5	−11	−8	−3	−6

1. ALGEBRA ______

2. INTEGER ______

3. POSITIVE ______

4. NEGATIVE ______

5. SIGN ______

6. ADDITION ______

7. GREATER ______

8. LESS ______

9. EQUAL ______

Write <, >, or = to compare the values of each word pair.

10. MANY ☐ FEW

11. ALL ☐ NOTHING

12. SCHOOL ☐ HOME

13. DOG ☐ CAT

14. ADD ☐ SUBTRACT

15. STOP ☐ GO

16. EVALUATE ☐ SOLVE

17. MORE ☐ LESS

18. EMPTY ☐ FULL

19. Write your name and the first names of 3 other friends or family members. Find the value of each name. Write the names in order from least to greatest value.

____

20. Write a word with a value of 3.

____

21. Write a word with a value of −3.

____

22. What is the highest value word you can think of? Compare it with your classmates' words.

____

Name ______________________ Date ___________ Class ___________

LESSON **2-2**

Problem Solving

Adding Integers

Write the correct answer.

1. The temperature dropped 12°F in 8 hours. If the final temperature was −7°F, what was the starting temperature?

____

2. At 3 P.M., the temperature was 9°F. By 11 P.M., it had dropped 31°F. What was the temperature at 11 P.M.?

____

3. Tad owes John $23 and borrows $12 more. How much does Tad owe John now?

____

4. New Orleans, Louisiana, is 6 feet below sea level. The highest point in Louisiana, Driskill Mountain, is 541 feet higher than New Orleans. How high is Driskill Mountain?

____

5. A submarine submerged at a depth of −40 ft. dives 57 ft more. What is the new depth of the submarine?

____

6. An airplane at 20,000 ft drops 2,500 ft in altitude. What is the new altitude?

____

Choose the letter for the best answer.

7. Last week, Jane made deposits of $64, $25, and $37 into her checking account. She then wrote checks for $52 and $49. What is the overall change in Jane's account balance?

A −$99 **C** $126

B $25 **D** $227

8. In Indianapolis, Indiana, the coldest recorded temperature was −23°F. The hottest recorded temperature was 127°F higher. What was the hottest temperature in Indianapolis?

F 150°F **H** 104°F

G 127°F **J** −150°F

9. Helena borrowed $189 from her parents to buy an electric bass. She paid back $56 last week and $64 this week. How much does Helena still owe her parents?

A $133 **C** $69

B $120 **D** $29

10. The Aral Sea and the Caspian Sea are actually lakes. The elevation of the Caspian Sea is 92 feet below sea level. The Aral Sea is 217 feet higher. What is the elevation of the Aral Sea?

F −125 ft **H** 309 ft

G −309 ft **J** 125 ft

Name ______________________ Date __________ Class __________

LESSON **2-2**

Reading Strategies

Use Graphic Aids

Randy's football team had the ball on the zero yard line. On their first play they gained six yards. On the second play they lost four yards. On what yard line is the ball now?

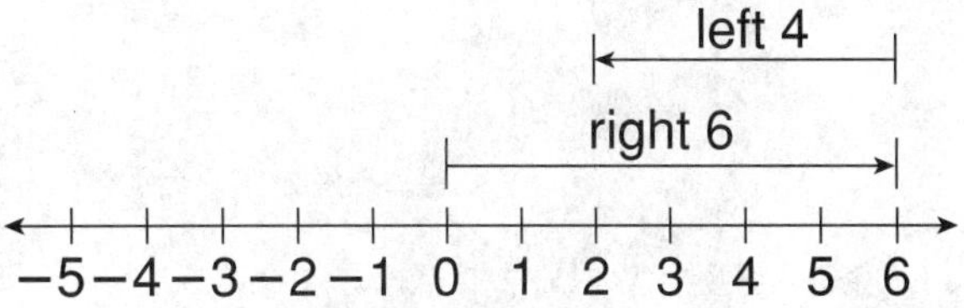

$6 + (-4) = 2$

Use the number line to help you answer the questions.

1. On which number do you begin? __________
2. Which direction do you move first? How many places do you move?

   ____
3. Which direction do you move next? How many places do you move?

   ____

When Angela went to bed, the temperature was zero degrees. When her mother went to bed two hours later, the temperature had gone down 5 degrees. By the time Angela got up the temperature had gone down another 3 degrees. What was the temperature when she got up?

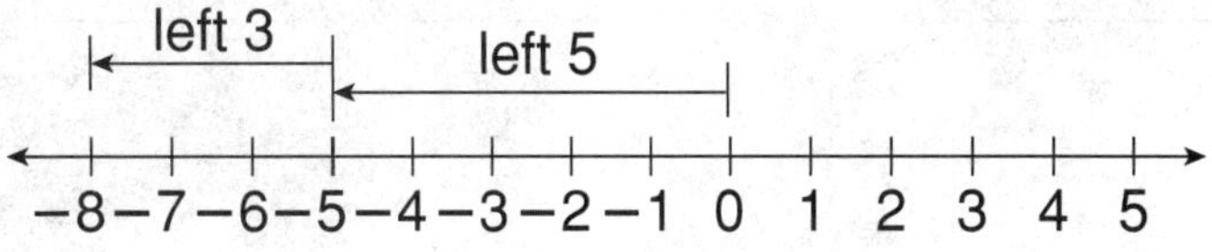

$-5 + (-3) = -8$

Use the number line to help you answer the questions.

4. On which number do you begin? __________
5. Which direction do you move first? How many spaces?

   ____
6. Which direction do you move next? How many spaces?

   ____

Name ______________________ Date ___________ Class ___________

LESSON **2-2**

Puzzles, Twisters & Teasers

Chill Out!

Add each pair of integers. Write the letter on the line above the correct answer at the bottom of the page to solve this riddle:

Why did the girl put her money in the freezer?

R $7 + (-16)$

S $-3 + 10$

L $-6 + (-13)$

H $-7 + (-81)$

O $-1 + (-5)$

E $2 + (-20)$

C $25 + 14$

W $-3 + 28$

D $9 + 4$

A $-5 + (-12)$

T $6 + (-22)$

N $8 + (-7)$

___ 7	___ −88	___ −18		___ 25	___ −17	___ 1	___ −16	___ −18	___ 13

___ 39	___ −6	___ −19	___ 13		___ −88	___ −17	___ −9	___ 13		___ 39	___ −17	___ 7	___ −88	!

Name ______________________ Date ____________ Class ____________

LESSON 2-3

Practice A

Subtracting Integers

Show the subtraction on the number line. Then write the difference.

1. $3 - 8$

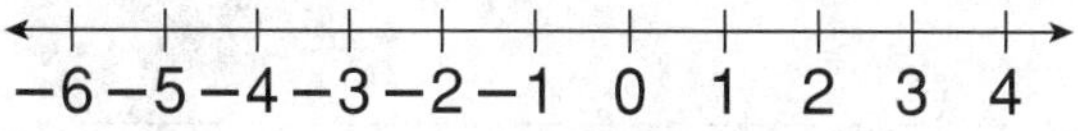

____

2. $-5 - (-1)$

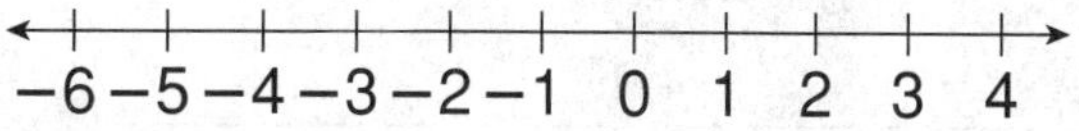

____

Find each difference.

3. $-3 - 4$ ________

4. $-7 - (-2)$ ________

5. $12 - 6$ ________

6. $2 - (-7)$ ________

7. $-8 - 8$ ________

8. $-5 - (-5)$ ________

9. $-1 - (-2)$ ________

10. $9 - (-3)$ ________

11. $8 - 1$ ________

12. $7 - (-9)$ ________

13. $-3 - 8$ ________

14. $-3 - (-7)$ ________

Evaluate $x - y$ for each set of values.

15. $x = 6$, $y = -3$ ________

16. $x = -7$, $y = 1$ ________

17. $x = -2$, $y = -5$ ________

18. $x = 9$, $y = 11$ ________

19. $x = -1$, $y = -1$ ________

20. $x = -5$, $y = 5$ ________

21. The high temperature one day was 6°F. The low temperature was −3°F. What was the difference between the high and low temperatures for the day?

____

22. The temperature changed from −7°F at 6 A.M. to 7°F at noon. How much did the temperature increase?

____

Name ______________________ Date ____________ Class ____________

LESSON **2-3**

Practice B

Subtracting Integers

Use a number line to find each difference.

1. $-2 - 3$

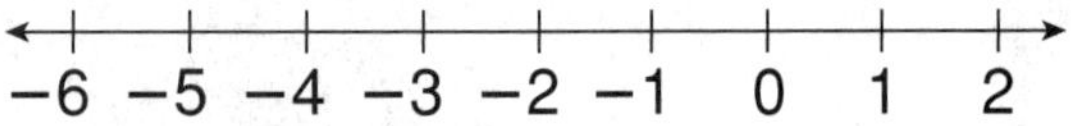

____

2. $5 - (-1)$

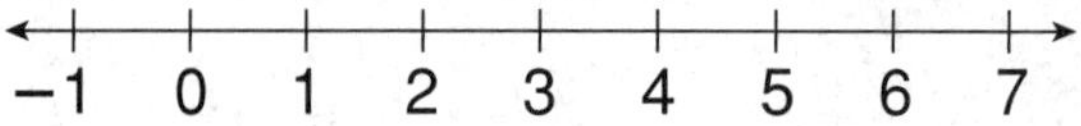

____

Find each difference.

3. $-6 - 4$ ____________

4. $-7 - (-12)$ ____________

5. $12 - 16$ ____________

6. $5 - (-19)$ ____________

7. $-18 - (-18)$ ____________

8. $23 - (-23)$ ____________

9. $-10 - (-9)$ ____________

10. $29 - (-13)$ ____________

11. $9 - 15$ ____________

12. $-12 - 14$ ____________

13. $22 - (-8)$ ____________

14. $-16 - (-11)$ ____________

Evaluate $x - y$ for each set of values.

15. $x = 14, y = -2$ ____________

16. $x = -11, y = 11$ ____________

17. $x = -8, y = -15$ ____________

18. $x = -9, y = -9$ ____________

19. $x = 19, y = -20$ ____________

20. $x = 20, y = 25$ ____________

21. The high temperature one day was −1°F. The low temperature was −5°F.What was the difference between the high and low temperatures for the day?

____

22. The temperature changed from 5°F at 6 P.M. to −2°F at midnight. How much did the temperature decrease?

____

Name ____________________ Date __________ Class __________

LESSON **2-3**

Practice C

Subtracting Integers

Use a number line to find each difference.

1. −6 − (−3)

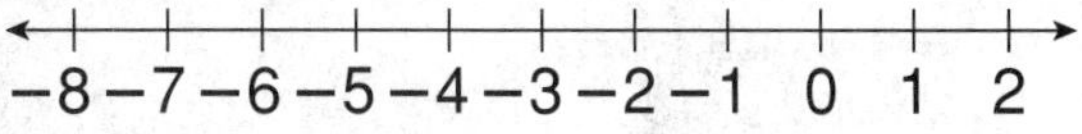

____

2. −1 − 5

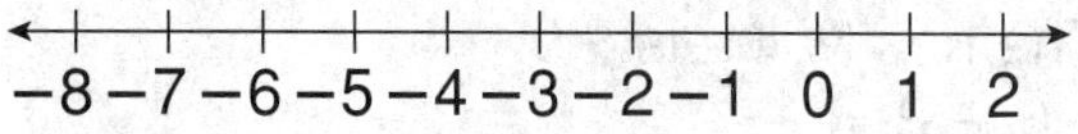

____

Find each difference.

3. −16 − 15 ________

4. −24 − (−12) ________

5. 25 − 41 ________

6. 19 − (−19) ________

7. 48 − (−18) ________

8. −37 − 37 ________

9. −50 − (−50) ________

10. −39 − 13 ________

Evaluate *x* − *y* for each set of values.

11. $x = -24$, $y = 24$ ________

12. $x = -22$, $y = -32$ ________

13. $x = 18$, $y = 25$ ________

Simplify.

14. 9 + (−15) − 6 ________

15. 12 − (−5) −3 ________

16. −13 − (−8) − (−5) ________

17. −4 −9 + (−7) ________

18. 10 + (−3) − (−18) ________

19. −25 − 37 − (−5) ________

20. The record high temperature in North Carolina is 110°F. The record low temperature is −34°F. What is the difference in these temperatures?

____

21. The temperature at 6 P.M. was 15°F. By 8 P.M. it had dropped 9 degrees, and then by 10 P.M. it had dropped another 11 degrees. What was the temperature at 10 P.M.?

____

Name ______________________________ Date __________ Class __________

LESSON **2-3**

Reteach

Subtracting Integers

The total value of the three cards shown is −6.

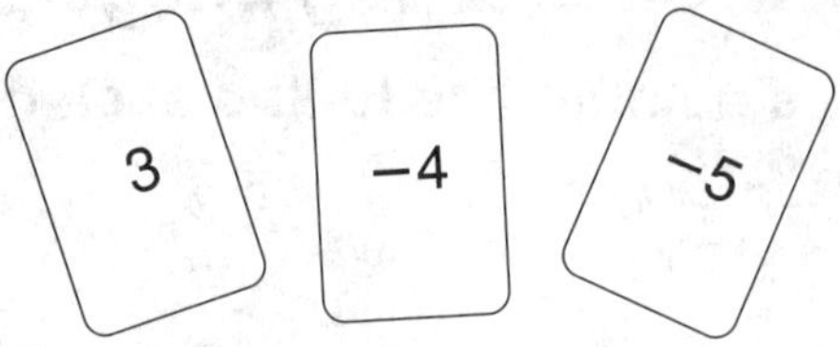

What if you **take away** the 3 card?
Cards −4 and −5 are left.
The new value is −9.
$-6 - 3 = -9$

What if you **take away** the −4 card?
Cards 3 and −5 are left.
The new value is −2.
$-6 - (-4) = -2$

Answer each question.

1. Suppose you have the cards shown.
The total value of the cards is 12.

a. What if you take away the 7 card? $12 - 7 =$ ____________________

b. What if you take away the 13 card? $12 - 13 =$ ____________________

c. What if you take away the −8 card? $12 - (-8) =$ ____________________

2. Subtract $-4 - (-2)$.

a. $-4 < -2$. Will the answer be positive or negative? ____________________

b. $|4| - |2| =$ ____________________

c. $-4 - (-2) =$ ____________________

3. Subtract $21 - 13$.

a. $21 > 13$. Will the answer be positive or negative? ____________________

b. $|21| - |13| =$ ____________________

c. $21 - 13 =$ ____________________

Subtract.

4. $31 - (-9) =$ ______

5. $15 - 18 =$ ______

6. $-9 - 17 =$ ______

7. $-8 - (-8) =$ ______

8. $29 - (-2) =$ ______

9. $13 - 18 =$ ______

Name ______________________ Date __________ Class __________

LESSON 2-3

Challenge

Subtract-a-Fact

Complete each temperature and elevation fact. Write the answer in the tables below.

Temperature Facts

1. Louisiana's lowest recorded temperature is (−15 − 1)°F.
2. The highest recorded temperature for Texas is [−18 − (−138)]°F.
3. The normal January temperature for Fairbanks, Alaska, is (−5 − 5)°F.
4. The normal July temperature for Tampa, Florida, is [45 − (−37)]°F.
5. The lowest temperature of 2000 in Detroit, Michigan, was (87 − 90)°F.
6. The highest temperature of 2000 in Norfolk, Virginia, was [−42 − (−138)]°F.
7. North Carolina's lowest recorded temperature is (57 − 91)°F.
8. South Carolina's lowest recorded temperature is (−4 − 15)°F.

Elevation Facts

9. Death Valley's elevation is (−25 − 257) ft.
10. The Caspian Sea's elevation is (125 − 217) ft.
11. Mount Everest's elevation is [−26 − (−29,061)] ft.
12. Mount McKinley's elevation is Mount Everest's elevation − 8,715 ft.
13. Lake Eyre's elevation is (68 − 120) ft.
14. The Dead Sea's elevation is (−761 − 587) ft.
15. Mount Elbrus's elevation is [11,500 − (−7,010)] ft.
16. Lake Assal's elevation is (−254 − 258) ft.

Temperature Facts	°F
Louisiana's lowest	
Texas's highest	
Fairbanks normal January	
Tampa normal July	
Lowest in Detroit in 2000	
Highest in Norfolk in 2000	
North Carolina's lowest	
South Carolina's lowest	

Highest and Lowest Points on Continents	Elevation
Death Valley, North America	
Caspian Sea, Europe	
Mount Everest, Asia	
Mount McKinley, North America	
Lake Eyre, Australia	
Dead Sea, Asia	
Mount Elbrus, Europe	
Lake Assal, Africa	

Name ______________________ Date ____________ Class ____________

LESSON 2-3

Problem Solving

Subtracting Integers

Write the correct answer.

1. The daytime temperature on Mercury can reach 430°C. The nighttime temperature can drop to −180°C. How much can the temperature drop during one day?

____

2. An ice cream company reported a net profit of $24,000 in 2002 and a net loss of $11,000 in 2003. How much did the company's profits change from 2002 to 2003?

____

3. A small business reported a net loss of $86,000 in 2004 and a net profit of $32,000 in 2005. How much did the company's profits change from 2004 to 2005?

____

4. The daytime high temperature on the Moon can reach 130°C. At night, the temperature can drop to −110°C. What is the difference between the high and low temperatures?

____

Choose the letter for the best answer.

5. The low point of the Tonga Trench, in the Pacific Ocean, is −10,630 meters. The low point of the Mariana Trench, also in the Pacific Ocean, is 890 meters lower. What is the depth of the Mariana Trench?

A 10,630 meters

B −11,520 meters

C −9,740 meters

D 9,740 meters

6. On Wednesday night in St. Petersburg, Russia, the temperature is −11°C. On the same night in Bombay, India, the temperature is 17°C. What is the difference in temperature?

F −6°C

G 50°C

H −187°C

J 28°C

7. Climax, Colorado, is the highest town in the United States at 11,560 feet. The lowest town is Calipatria, California, which is 185 feet below sea level. What is the difference in elevation?

A −185 feet

B 11,375 feet

C 11,560 feet

D 11,745 feet

8. The low point of the Japanese Trench, in the Pacific Ocean, is −10,372 meters. The low point of the Puerto Rico Trench, in the Atlantic Ocean, is 1,172 meters higher. What is the depth of the Puerto Rico Trench?

F −9,200 meters

G 8,200 meters

H −1,172 meters

J −11,544 meters

Name ______________________ Date ____________ Class ____________

LESSON 2-3

Reading Strategies

Use Graphic Aids

Brett borrowed $7 from his father to buy a CD. He paid back $3. How much money does Brett have now? The number line will help you picture this problem.

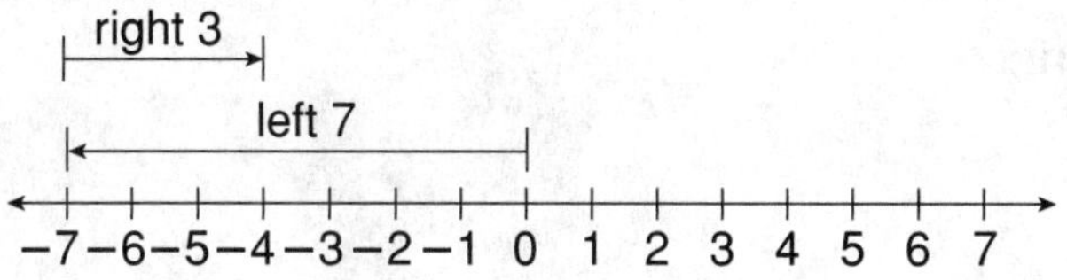

$3 - 7 = -4$

1. Beginning at 0, in which direction will you move first? ____________

2. How many places? ____________

3. Which direction do you move next? ____________

4. How many places? ____________

5. On what number do you end? ____________

Bret does not have any money. He owes his dad $4. He has negative $4.

Sally and her friends made up a game with points. You can either win or lose up to ten points on each round of the game. After the first round, Sally's team had 2 points. In the second round they lost 6 points. How many points was Sally's team down by after the second round?

The number line will help you picture the problem.

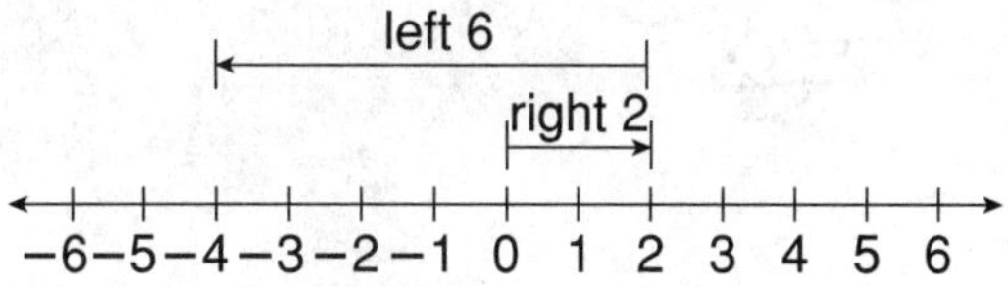

$2 - 6 = -4$

6. Beginning at zero, which direction will you move first? How many places?

____

7. Which direction will you move next? How many places?

____

8. By how many points was Sally's team down? ____________

Name ______________________ Date ____________ Class ____________

LESSON 2-3

Puzzles, Twisters & Teasers

Integer Invasion!

Subtract each pair of integers. Write the letter on the line above the correct answer at the bottom of the page to answer the following question:

What do you call a pan spinning through space?

C $-12 - (-2)$

A $-1 - 8$

J $-3 - (-8)$

N $6 - 10$

B $-31 - 15$

U $12 - (-6)$

O $-15 - 15$

I $30 - (-20)$

G $-22 - (-5)$

D $-4 - 21$

Y $27 - 19$

E $-10 - (-7)$

R $2 - 9$

T $-5 - 9$

F $-8 - (-2)$

___	___
−9	−4

___	___	___	___	___	___	___	___	___	___	___	___
18	−4	50	−25	−3	−4	−14	50	−6	50	−3	−25

___	___	___	___	___	___		___	___	___	___	___	___
−6	−7	8	50	−4	−17		−30	−46	5	−3	−10	−14

Name ________________ Date __________ Class __________

LESSON **2-4**

Practice A

Multiplying and Dividing Integers

Find each product.

1. $6 \cdot (-1)$ ________

2. $-4 \cdot 2$ ________

3. $-3 \cdot (-4)$ ________

4. $-2 \cdot 8$ ________

5. $5 \cdot (-7)$ ________

6. $-7 \cdot 9$ ________

7. $8 \cdot 4$ ________

8. $-3 \cdot (-5)$ ________

Multiply.

9. $-5 \cdot (-5)$ ________

10. $8 \cdot (-4)$ ________

11. $-7 \cdot (-6)$ ________

12. $9 \cdot (-8)$ ________

13. $1 \cdot (-7)$ ________

14. $-4 \cdot (-5)$ ________

15. $-6 \cdot 3$ ________

16. $-7 \cdot (-7)$ ________

Find each quotient.

17. $12 \div (-4)$ ________

18. $-15 \div (-3)$ ________

19. $-20 \div 5$ ________

20. $-27 \div (-9)$ ________

21. $-45 \div (-5)$ ________

22. $-18 \div 9$ ________

23. $24 \div (-4)$ ________

24. $32 \div 4$ ________

25. $21 \div 3$ ________

26. $-36 \div (-4)$ ________

27. $16 \div (-4)$ ________

28. $-56 \div 8$ ________

29. $-42 \div 7$ ________

30. $-30 \div (-6)$ ________

31. $27 \div 9$ ________

32. $25 \div (-5)$ ________

33. Jenny was walking down a wooded path. For 4 minutes, the elevation dropped steadily. Altogether it dropped 12 feet. What was the change in elevation during the first minute?

____

34. As a cold front passed, the temperature dropped steadily 2 degrees each hour. It dropped a total of 10 degrees. How many hours did it take for the temperature to drop this far?

____

Name ______________________ Date __________ Class __________

LESSON **2-4**

Practice B

Multiplying and Dividing Integers

Find each product.

1. $8 \cdot (-5)$ ________

2. $-4 \cdot 7$ ________

3. $-6 \cdot (-3)$ ________

4. $-2 \cdot 4$ ________

5. $4 \cdot (-9)$ ________

6. $-9 \cdot 5$ ________

7. $6 \cdot 8$ ________

8. $-7 \cdot (-3)$ ________

Multiply.

9. $-6 \cdot (-6)$ ________

10. $9 \cdot (-3)$ ________

11. $-2 \cdot (-8)$ ________

12. $5 \cdot (-7)$ ________

13. $10 \cdot 8$ ________

14. $-5 \cdot 9$ ________

15. $9 \cdot (-6)$ ________

16. $(-4) \cdot (-11)$ ________

Find each quotient.

17. $25 \div (-5)$ ________

18. $-54 \div (-6)$ ________

19. $-10 \div 5$ ________

20. $-28 \div (-4)$ ________

21. $-42 \div (-7)$ ________

22. $-21 \div 3$ ________

23. $36 \div (-6)$ ________

24. $-81 \div (-9)$ ________

25. $-32 \div 8$ ________

26. $45 \div (-9)$ ________

27. $-72 \div (-8)$ ________

28. $50 \div 10$ ________

29. $-42 \div 6$ ________

30. $-72 \div (-9)$ ________

31. $40 \div 8$ ________

32. $56 \div (-7)$ ________

33. Kim was walking down a rocky path. For 4 minutes, the elevation dropped steadily. Altogether it dropped 8 feet. What was the change in elevation per minute for the 4 minutes?

____

34. As a front passed, the temperature changed steadily over 6 hours. Altogether it changed -18 degrees. What was the change in temperature per hour for the 6 hours?

____

Name ______________________ Date __________ Class __________

LESSON 2-4

Practice C
Multiplying and Dividing Integers

Find each product.

1. $9 \cdot (-8)$ ________ **2.** $-7 \cdot 7$ ________ **3.** $-5 \cdot (-6)$ ________ **4.** $-4 \cdot 8$ ________

5. $12 \cdot (-10)$ ________ **6.** $-50 \cdot 3$ ________ **7.** $8 \cdot 10$ ________ **8.** $-20 \cdot (-6)$ ________

Multiply.

9. $-7 \cdot (-5)$ ________ **10.** $10 \cdot (-9)$ ________ **11.** $-8 \cdot (-8)$ ________ **12.** $12 \cdot (-3)$ ________

13. $-9 \cdot 8$ ________ **14.** $15 \cdot (-4)$ ________ **15.** $-5 \cdot (-25)$ ________ **16.** $-150 \cdot 2$ ________

Find each quotient.

17. $63 \div (-7)$ ________ **18.** $-48 \div (-4)$ ________ **19.** $-36 \div 9$ ________ **20.** $-81 \div (-9)$ ________

21. $-56 \div (-7)$ ________ **22.** $-54 \div 9$ ________ **23.** $80 \div (-10)$ ________ **24.** $-60 \div (-12)$ ________

25. $33 \div (-3)$ ________ **26.** $100 \div (-20)$ ________ **27.** $-75 \div (-3)$ ________ **28.** $96 \div 8$ ________

29. $-200 \div 50$ ________ **30.** $-80 \div (-4)$ ________ **31.** $48 \div 12$ ________ **32.** $77 \div (-11)$ ________

33. As Randy swam underwater, his elevation changed steadily for 4 minutes. Altogether it changed -20 feet. What was the change in elevation during each minute?

____

34. Marie withdrew the same amount of money from her savings account each month for 5 months. Altogether she withdrew \$300. She did not make any deposits. What was the change in Marie's savings account balance each month?

____

Name ______________________ Date __________ Class __________

LESSON 2-4

Reteach

Multiplying and Dividing Integers

Look for the patterns in these products and quotients.

$1 \cdot 3 = 3$	$-1 \cdot 3 = -3$	$3 \div 1 = 3$	$3 \div (-1) = -3$
$2 \cdot 3 = 6$	$-2 \cdot 3 = -6$	$6 \div 2 = 3$	$6 \div (-2) = -3$
$-3 \cdot (-3) = 9$	$3 \cdot (-3) = -9$	$-9 \div (-3) = 3$	$-9 \div 3 = -3$
$-4 \cdot (-3) = 12$	$4 \cdot (-3) = -12$	$-12 \div (-4) = 3$	$-12 \div 4 = -3$

Look at how to find the signs of the products.

- The product of two integers with the **same sign** is **positive.**

$(+) \cdot (+) = (+) \qquad (-) \cdot (-) = (+)$

- The product of two integers with **different signs** is **negative.**

$(+) \cdot (-) = (-) \qquad (-) \cdot (+) = (-)$

Look at how to find the signs of the quotients.

- The quotient of two integers with the **same sign** is **positive.**

$(+) \div (+) = (+) \qquad (-) \div (-) = (+)$

- The quotient of two integers with **different signs** is **negative.**

$(+) \div (-) = (-) \qquad (-) \div (+) = (-)$

Find each product or quotient.

1. $-5 \cdot 4$ ________

2. $2 \cdot (-8)$ ________

3. $-1 \cdot (-1)$ ________

4. $-6 \cdot 3$ ________

5. $7 \cdot (-3)$ ________

6. $-8 \cdot (-4)$ ________

7. $-6 \cdot 5$ ________

8. $-9 \cdot (-9)$ ________

9. $36 \div (-4)$ ________

10. $-27 \div 9$ ________

11. $-24 \div (-6)$ ________

12. $-30 \div 5$ ________

13. $18 \div 6$ ________

14. $32 \div (-8)$ ________

15. $-45 \div 9$ ________

16. $-40 \div (-10)$ ________

Name ______________________________ Date __________ Class __________

LESSON **2-4**

Challenge

Football Follies

Phil the Fumbler and Sack'im Sam play for two different football teams. Last year, they both had their worst season ever. Read their interview with a local sports announcer. Find out which player had the worse year by completing the table below.

Announcer: Well Phil, you didn't have a very good season, did you?

Phil: No, I guess losing 42 yards in each of the first 6 games was not what I had hoped to do.

Sam: Gee, that wasn't a very good opening, Phil. At least I gained 53 yards in each of my first 2 games, although, I did lose 88 yards in each of the next 4 games.

Announcer: Yes, Sam, that's true. Now can you tell us how you did in games 7 through 12 last season?

Sam: I can, but I don't really want to.

Announcer: I heard you were sacked and lost 125 yards in each of those games. Is this accurate.

Sam: Yup.

Announcer: Phil, how did you do in games 7 through 12?

Phil: Not that bad. I fumbled quite a bit in games 7 through 11, so I lost 130 yards in each of those games. But I came back in game 12 and gained 85 yards during that one.

Announcer: And what was your yardage for games 13 through 16?

Phil: I lost 95 yards in each game.

Sam: Me, too.

Announcer: Well, I sure hope next season is better for both of you.

Phil and Sam: ME, TOO!

Complete the table.

	Phil	**Sam**
Total Yards Gained		
Total Yards Lost		
Total Yards		
Average per Game for 16 Games (to the nearest tenth)		
Which player had the worse year?		

Name ______________________ Date __________ Class __________

LESSON 2-4

Problem Solving

Multiplying and Dividing Integers

Gerald recorded the temperature at 6-hour intervals for one 24-hour period. The chart shows his data.

Time	10:00 P.M.	4:00 A.M.	10:00 A.M.	4:00 P.M.	10:00 P.M.
Temperature (°F)	70	58	64	82	76

Write the correct answer.

1. If the temperature changed steadily, what was the change per hour from 10:00 P.M. the first day to 4:00 A.M.?

____

2. If the temperature changed steadily, what was the change per hour from 4:00 A.M. to 10:00 A.M.?

____

3. If the temperature changed steadily, what was the change per hour from 10:00 A.M. to 4:00 P.M.?

____

4. If the temperature changed steadily, what was the change per hour from 4:00 P.M. to 10:00 P.M. the second day?

____

Choose the letter for the best answer.

5. A small company had a profit of −$528 in January. If it continues to have the same profit each month for 4 months, what will be the company's total profit for 4 months?

A −$132 **C** −$2,112

B $132 **D** $2,112

6. Evi's watch is water resistant up to −15 feet. Mateo's watch is water resistant up to 8 times the depth of Evi's watch. Mateo's watch is water resistant to what maximum depth?

F −7 feet **H** −120 feet

G 120 feet **J** 7 feet

7. A submarine at −235 meters dives to a depth 8 times its initial depth. To what depth does the submarine dive?

A −227 meters

B −243 meters

C −1,880 meters

D −8,235 meters

8. A 1-kilogram rock dropped into the ocean would take 37 minutes to reach −6,660 meters. If the rock dropped steadily, how far would it fall in 1 minute?

F 167 meters

G −167 meters

H −180 meters

J 188 meters

Name ______________________ Date __________ Class __________

LESSON **2-4**

Reading Strategies

Use Graphic Aids

The opposite of 5 is negative 5. Owing money is the opposite of having money. Owing $5 is the opposite of having $5.

1. What is the opposite of owing $10? ____________________

2. What is the opposite of having $17? ____________________

David borrowed $4 from his mother each of the last three months. How much money does he owe his mother? The money he owes his mother is a negative number. This problem can be pictured on a number line.

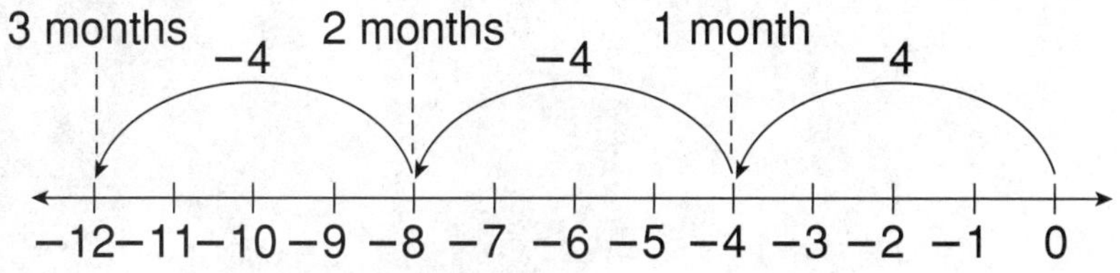

$3 \cdot (-4) = -12$

Use the number line to help you answer the questions.

3. Starting at zero, which direction do you move first? ____________________

4. How many places do you move? ____________________

5. Which direction do you move next? ____________________

6. How many places do you move? ____________________

7. Which direction do you move next? ____________________

8. How many places do you move? ____________________

9. How much money does David owe his mother? ____________________

10. If David borrowed $4 for one more month, how much would he owe his mother? ____________________

Name ____________________ Date __________ Class __________

LESSON 2-4 Puzzles, Twisters & Teasers

It's Raining Money!

Solve the equations. Then use the letter associated with the answer to solve the riddle.

H $3 \cdot (-3) =$ ______

E $4 \cdot 2 =$ ______

N $4 \cdot (-2) =$ ______

T $-4 \cdot 3 =$ ______

I $3 \cdot 3 =$ ______

S $2 \cdot (-3) =$ ______

C $-4 \cdot 2 =$ ______

G $3 \cdot 2 =$ ______

A $-3 \cdot 2 =$ ______

R $3 \cdot 2 =$ ______

When does it rain money?

W ___ ___ ___ ___ ___ ___ ___ ___

−9 8 −8 −12 −9 8 10 8

___ ___ ___ ___ ___ ___ ___ ___ ___ ___

9 −6 −8 −9 −6 −8 6 8 9 −8

___ ___ ___ **W** ___ ___ ___ ___ ___ ___

−12 −9 8 8 −6 −12 −9 8 10

Name ______________________ Date ____________ Class ____________

LESSON **2-5**

Practice A

Solving Equations Containing Integers

Solve. Check each answer.

1. $n - 6 = -2$ ____________

2. $x - 8 = -11$ ____________

3. $7 = a - 5$ ____________

4. $y + 4 = 2$ ____________

5. $c + 7 = -3$ ____________

6. $0 = v + 1$ ____________

7. $8j = -16$ ____________

8. $-3k = 24$ ____________

9. $-20 = -4s$ ____________

10. $\frac{m}{-2} = -5$ ____________

11. $\frac{d}{6} = -3$ ____________

12. $\frac{r}{-7} = 4$ ____________

13. $p + 8 = -6$ ____________

14. $-15 = 5b$ ____________

15. $f - 9 = -1$ ____________

16. $\frac{n}{6} = -4$ ____________

17. $k + 10 = 3$ ____________

18. $4a = -16$ ____________

19. $-6x = -36$ ____________

20. $2 = e - 7$ ____________

21. $3 = \frac{m}{2}$ ____________

22. The temperature in Minnesota was −8°F one day. This was 12 degrees less than the temperature in Indiana on the same day. What was the temperature in Indiana?

____

23. Mr. Harding sold 100 shares of stock at $14 per share. He had a loss of $6 per share. What did Mr. Harding pay for each share of the stock?

____

Name ______________________ Date __________ Class __________

LESSON 2-5

Practice B

Solving Equations Containing Integers

Solve. Check each answer.

1. $y - 5 = -4$

2. $n - 9 = -14$

3. $13 = x - 15$

4. $p + 18 = 14$

5. $q + 6 = -2$

6. $0 = w + 4$

7. $9h = -36$

8. $-3b = 36$

9. $-100 = -4u$

10. $\frac{d}{5} = -7$

11. $\frac{c}{4} = -20$

12. $\frac{s}{-9} = 9$

13. $f + 15 = -16$

14. $-75 = 3v$

15. $g - 19 = -21$

16. $-63 = -9s$

17. $14 + m = -10$

18. $12 = \frac{w}{4}$

19. $x = 15 - 31$

20. $\frac{e}{-7} = 8$

21. $-6 = 21 - n$

22. The temperature in Buffalo, New York, was −2°F one day. This was 42 degrees warmer than the temperature in Nome, Alaska, on the same day. What was the temperature in Nome?

23. LaSanda bought 20 shares of stock for $175. She sold the stock for a total profit of $25. What was the selling price of each share of stock?

Name ______________________ Date __________ Class __________

LESSON 2-5

Practice C

Solving Equations Containing Integers

Solve. Check each answer.

1. $h - 15 = -17$ ______

2. $a - 29 = -12$ ______

3. $13 = g - 20$ ______

4. $m + 48 = 31$ ______

5. $n + 26 = -4$ ______

6. $0 = z + 50$ ______

7. $15y = -75$ ______

8. $0 = -7d$ ______

9. $-144 = -8v$ ______

10. $\frac{k}{-5} = -14$ ______

11. $\frac{r}{4} = -250$ ______

12. $\frac{t}{-11} = 11$ ______

13. $s + 25 = -18$ ______

14. $-96 = -4j$ ______

15. $x - 49 = -57$ ______

Translate each sentence into an equation. Then solve the equation.

16. 8 less than a number *n* is –13.

______

17. A number *f* multiplied by –12 is –72.

______

18. The sum of –47 and a number *d* is 84.

______

19. 8 less than a number *n* is –13.

______

20. Jake bought 50 shares of stock for $1,000 and sold them for a loss of $375. What was the selling price of each share of stock?

______

21. Mrs. Lee bought 50 shares of Wiget stock. She paid a total of $1,200, which included a $25 commission fee. Two years later she sold all 50 shares for a total of $1,500. This included a $25 commission fee to the broker. What was the cost and selling price per share of the stock?

______

Name ______________________ Date __________ Class __________

LESSON 2-5

Reteach

Solving Equations Containing Integers

- You can use addition to solve an equation involving subtraction. Addition undoes subtraction. Adding the same number to both sides of the equation keeps the equation balanced.

	Check
$x - 5 = -6$	$x - 5 = -6$
$x - 5 + 5 = -6 + 5$	$-1 - 5 \stackrel{?}{=} -6$
$x = -1$	$-6 \stackrel{?}{=} -6$ ✓

- You can use subtraction to solve an equation involving addition. Subtraction undoes addition. Subtracting the same number from both sides of the equation keeps the equation balanced.

	Check
$n + 4 = -15$	$n + 4 = -15$
$n + 4 - 4 = -15 - 4$	$-19 + 4 \stackrel{?}{=} -15$
$n = -19$	$-15 \stackrel{?}{=} -15$ ✓

Solve. Check your answer.

1. $p - 9 = -3$

$p - 9 +$ ____ $= -3 +$ ____

____

2. $w - 2 = -14$

$w - 2 +$ ____ $= -14 +$ ____

____

3. $x - 12 = -5$

$x - 12 +$ ____ $= -5 +$ ____

____

4. $f - 8 = 6$

$f - 8 +$ ____ $= 6 +$ ____

____

5. $6 = m - 7$ __________

6. $-4 = s - 10$ __________

7. $-8 = y - 2$ __________

8. $a + 19 = 7$ __________

9. $b + 15 = -9$ __________

10. $39 + t = 45$ __________

11. $-5 = x + 7$ __________

12. $-2 = k + 11$ __________

13. $10 = -3 + j$ __________

Name ______________________ Date __________ Class __________

LESSON 2-5

Reteach

Solving Equations Containing Integers (continued)

- You can use division to solve an equation involving multiplication.

  Division undoes multiplication. Dividing both sides of the equation by the same number keeps the equation balanced.

	Check
$3y = -9$	$3y = -9$
$\frac{3y}{3} = \frac{-9}{3}$	$3 \cdot (-3) \stackrel{?}{=} -9$
$y = -3$	$-9 \stackrel{?}{=} -9$ ✓

- You can use multiplication to solve an equation involving division.

  Multiplication undoes division. Multiplying both sides of an equation by the same number keeps the equation balanced.

	Check
$\frac{a}{-5} = -8$	$\frac{a}{-5} = -8$
$-5 \cdot \frac{a}{-5} = -8 \cdot (-5)$	$\frac{40}{-5} \stackrel{?}{=} -8$
$a = 40$	$-8 \stackrel{?}{=} -8$ ✓

Solve. Check your answer.

14. $5g = -35$

$\frac{5g}{\_\_\_\_} = \frac{-35}{\_\_\_\_}$

____

15. $-8y = -96$

$\frac{-8y}{\_\_\_\_} = \frac{-96}{\_\_\_\_}$

____

16. $54 = -6f$

$\frac{54}{\_\_\_\_} = \frac{-6f}{\_\_\_\_}$

____

17. $3e = -33$

____

18. $-49 = 7n$

____

19. $-75 = -5c$

____

20. $\frac{n}{4} = -15$

____

21. $\frac{m}{-6} = -9$

____

22. $\frac{s}{-10} = 8$

____

23. $4 = \frac{w}{-6}$

____

24. $9 = \frac{z}{5}$

____

25. $-11 = \frac{h}{6}$

____

Name ______________________________ Date ____________ Class ____________

LESSON **2-5**

Challenge

Balloon Trail

Match each statement with one of the equations A through J. Write the letter of the equation in the space provided. Then solve each equation and write the solution. To find the way to the party, draw a path through the balloons that show the correct solutions to Exercises 1 to 10 in order.

A. $-7 + x = -3$ **B.** $\frac{n}{4} = -8$ **C.** $3y = -36$ **D.** $x - 9 = -4$

E. $-6n = 42$ **F.** $\frac{y}{5} = 10$ **G.** $x + 12 = -5$ **H.** $n - 6 = 2$

I. $-9y = -27$ **J.** $\frac{y}{-8} = -2$

1. Three times a number is -36. ______________

2. The sum of a number and 12 is -5. ______________

3. A number decreased by 9 is -4. ______________

4. A number divided by 4 is -8. ______________

5. A number divided by 5 is 10. ______________

6. A number divided by -8 is -2. ______________

7. 6 less than a number is 2. ______________

8. The product of a number and -6 is 42. ______________

9. Add -7 to a number. The sum is -3. ______________

10. -9 times a number is -27. ______________

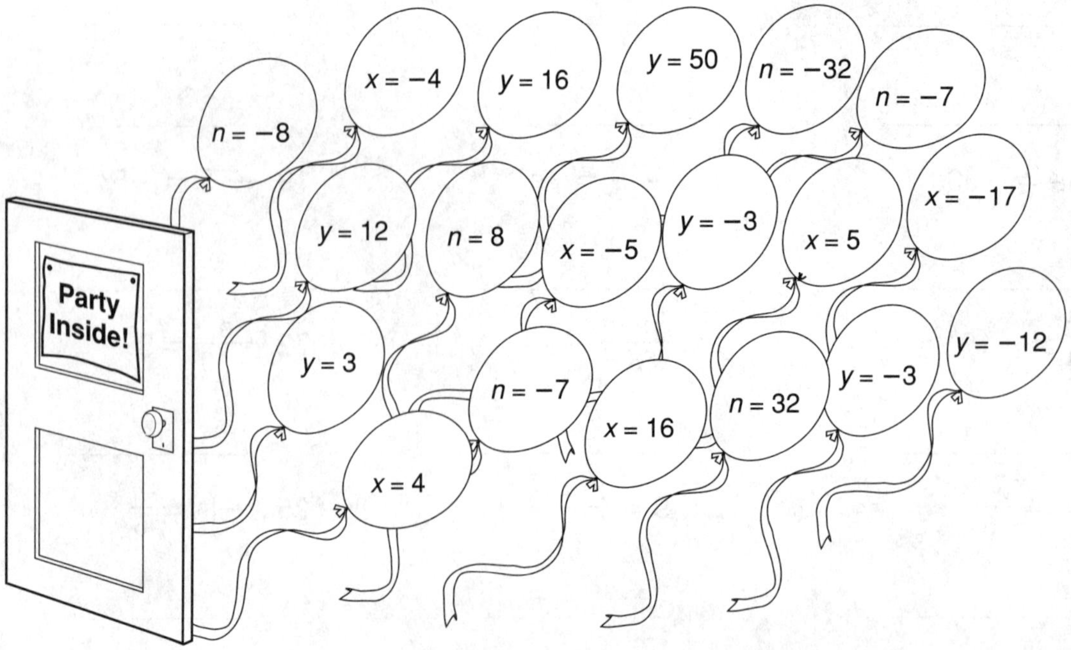

Name ______________________________ Date __________ Class __________

LESSON **2-5**

Problem Solving

Solving Equations Containing Integers

Write the correct answer.

1. Jolene has 30 days to prepare for a bicycle race. She will bicycle 15 miles each day. How many miles will Jolene have bicycled during her training?

____

2. When the amount of money spent is greater than the amount of income, it is called a deficit. In 2002, the U.S. budget deficit was −$158 billion. By 2004, it was −$412 billion. How much did the deficit go up during those two years?

____

3. Two angles of a triangle have a sum of 110°. The sum of all three angles is 180°. What is the measure of the third angle?

____

____

4. The Statue of Liberty was erected in New York City in 1886. How old is the statue?

____

____

Choose the letter for the best answer.

5. Lonnie swam 3 miles each day for 45 days. How many miles did Lonnie swim?

A 135 miles

B 48 miles

C 125 miles

D 45 miles

6. On the hottest day in Richmond, Virginia, the temperature was 105°F. On the coldest day, the temperature was 117°F lower. What was the coldest temperature in Richmond?

F −117°F

G −12°F

H 12°F

J 117°F

7. Mr. Marco sold 50 shares of Gizmo stock for $1,250. What was the selling price of the stock per share?

A $1,200

B $250

C $50

D $25

8. The U.S. Bullion Depository at Fort Knox contains about 147 million troy ounces of gold. At a market price of $400 per troy ounce, what is the approximate value of the gold stored at Fort Knox?

F $6 billion

G $60 billion

H $560 million

J $560 billion

Name ______________________ Date __________ Class __________

LESSON **2-5**

Reading Strategies

Use a Flowchart

The rules for solving equations with integers are the same as with whole numbers.

Use a flowchart to help you follow the rules.

1		2		3		4
Get the variable by itself.	→	Perform the same operation on both sides of the equation.	→	Apply rules for computing integers.	→	Check the solution.

$x + 12 = 5$ → Get x by itself.

$x + 12 - \mathbf{12} = 5 - \mathbf{12}$ → Subtract 12 from both sides.

$x = \mathbf{5 + (-12)}$ → Add the opposite.

$x = -7$ → Check.

$x + 12 = 5$

$-7 + 12 \stackrel{?}{=} 5$

$5 = 5$

Use $w - 12 = (-4)$ to answer Exercises 1–4.

1. What operation is used in this equation?

____

2. What operation will you use to get the variable by itself?

____

3. Apply this operation to both sides of the equation.

____

4. What is the value of w?

____

Use $x + (-9) = (-4)$ to answer Exercises 5 and 6.

5. What operation is used in this problem?

____

6. What operation will you use to get x by itself?

____

Name ______________________________ Date __________ Class __________

LESSON **2-5**

Puzzles, Twisters & Teasers

A Bold Conclusion!

Solve each equation. Then use the boldfaced letters to answer the riddle.

Q $9y = 900$ ____________________

A $d - 15 = 45$ ____________________

U $j + 56 = -7$ ____________________

C $v - 39 = -16$ ____________________

F $w + 41 = 0$ ____________________

K $-85 = -5c$ ____________________

X $11y = -121$ ____________________

E $-23 = x + 35$ ____________________

O $200 = -25p$ ____________________

R $-20 = 2z$ ____________________

B $a - 25 = -5$ ____________________

S $s + 2 = -4$ ____________________

W $-5a = 300$ ____________________

D $k = -18 + 11$ ____________________

G $b - 7 = -16$ ____________________

What do you call ten ducks in a cardboard box?

___		___	___	___		___	___
60		20	−8	−11		−8	−41
___	___	___	___	___	___	___	___
100	−63	60	23	17	−58	−10	−6

Name ______________________ Date __________ Class __________

LESSON 2-6

Practice A

Prime Factorization

Tell whether each number is prime or composite.

1. 25 ____________

2. 16 ____________

3. 21 ____________

4. 29 ____________

Choose the letter for the best answer.

5. What is the prime factorization of 24?

A $2^3 \cdot 3$

B 3

C $2^2 \cdot 3^2$

D $2^3 \cdot 3^2$

6. What is the prime factorization of 50?

F $3 \cdot 5^2$

G $3^2 \cdot 5$

H $2 \cdot 5^2$

J

7. What is the prime factorization of 35?

A $2 \cdot 3 \cdot 5$

B $5 \cdot 7$

C $3^2 \cdot 5$

D $3^2 \cdot 7$

8. What is the prime factorization of 27?

F 3^3

G

H 9^2

J $2^2 \cdot 3^2$

Use a factor tree to find the prime factorization.

9. 25 ____________

10. 20 ____________

11. 32 ____________

12. 49 ____________

13. 30 ____________

14. 44 ____________

15. 63 ____________

16. 28 ____________

Use a step diagram to find the prime factorization.

17. 93 ____________

18. 110 ____________

19. 80 ____________

20. 42 ____________

21. 625 ____________

22. 98 ____________

23. 140 ____________

24. 92 ____________

Name ______________________ Date __________ Class __________

LESSON 2-6

Practice B

Prime Factorization

Tell whether each number is prime or composite.

1. 33 ______

2. 41 ______

3. 52 ______

4. 79 ______

5. 96 ______

6. 121 ______

7. 83 ______

8. 119 ______

Write the prime factorization of each number.

9. 57 ______

10. 49 ______

11. 88 ______

12. 95 ______

13. 105 ______

14. 98 ______

15. 52 ______

16. 42 ______

17. 68 ______

18. 91 ______

18. 60 ______

20. 72 ______

21. 56 ______

22. 144 ______

23. 370 ______

24. 168 ______

25. 124 ______

26. 515 ______

27. 725 ______

28. 220 ______

29. 450 ______

30. 1,000 ______

31. 1,040 ______

32. 2,500 ______

33. The prime factorization of a number is $3^2 \cdot 5 \cdot 11$. What is the number?

______

Name ______________________ Date __________ Class __________

LESSON 2-6

Practice C

Prime Factorization

Tell whether each number is prime or composite.

1. 91 ________

2. 117 ________

3. 109 ________

4. 73 ________

Write the prime factorization of each number.

5. 84 ________

6. 343 ________

7. 135 ________

8. 180 ________

Use a step diagram to find the prime factorization.

9. 290 ________

10. 900 ________

11. 575 ________

12. 90 ________

13. 220 ________

14. 350 ________

15. 495 ________

16. 480 ________

17. 2,000 ________

18. 1,040 ________

19. 1,650 ________

20. 3,000 ________

21. The prime factorization of a number is $2^2 \cdot 3^5$. What is the number?

____

22. You can factor 150 as 15 • 10. List four other ways can you factor 150.

____

23. A number *x* is a prime factor of 65 and 104. What is the number?

____

Write the composite number for each prime factorization.

24. $2^5 \cdot 5^2$ ________

25. $3^3 \cdot 5 \cdot 11$ ________

26. $4^3 \cdot 7 \cdot 13$ ________

27. $2^8 \cdot 3^5$ ________

28. The prime factors of a number are all the prime numbers between 20 and 30. No factor is repeated. What is the number?

____

Name ________________________ Date __________ Class __________

LESSON **2-6**

Reteach

Prime Factorization

To write the **prime factorization** of a number, write the number as the product of only prime numbers. A prime number has only two factors, itself and 1.

Use a *factor tree* to find the prime factorization of 18.

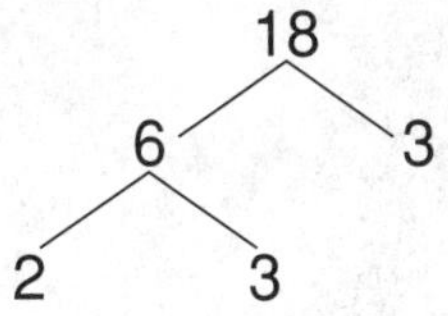

Keep factoring until all the factors are prime.

Use a *factor tree* to find the prime factorization of 36.

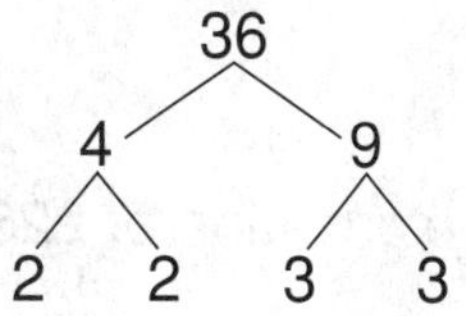

The prime factorization of 18 is $2 \cdot 3 \cdot 3$, or $2 \cdot 3^2$.

The prime factorization of 36 is $2 \cdot 2 \cdot 3 \cdot 3$, or $2^2 \cdot 3^2$.

Use a factor tree to find the prime factorization.

1. 20 ________ **2.** 28 ________ **3.** 54 ________ **4.** 63 ________

Use a *step diagram* to find the prime factorization of 60.

2|60
2|30
3|15
5|5
1

1) The divisors must be prime numbers.
2) Keep dividing until the quotient is 1.
3) The divisors are the factors in the prime factorization.

Use a *step diagram* to find the prime factorization of 75.

3|75
5|25
5|5
1

The prime factorization of 60 is $2 \cdot 2 \cdot 3 \cdot 5$, or $2^2 \cdot 3 \cdot 5$.

The prime factorization of 75 is $3 \cdot 5 \cdot 5$, or $3 \cdot 5^2$.

Use a step diagram to find the prime factorization.

5. 48 ________ **6.** 24 ________ **7.** 40 ________ **8.** 98 ________

Name ______________________ Date __________ Class __________

LESSON 2-6

Challenge

A Different Point of View

Create two more factor trees with different factors for each number.

1.

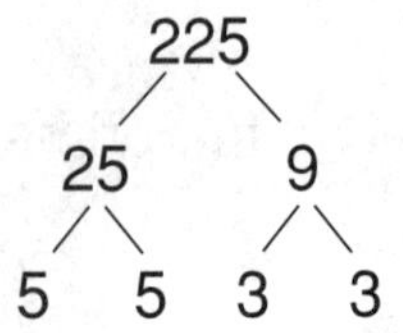

225

225

Prime factorization of 225: __________

2.

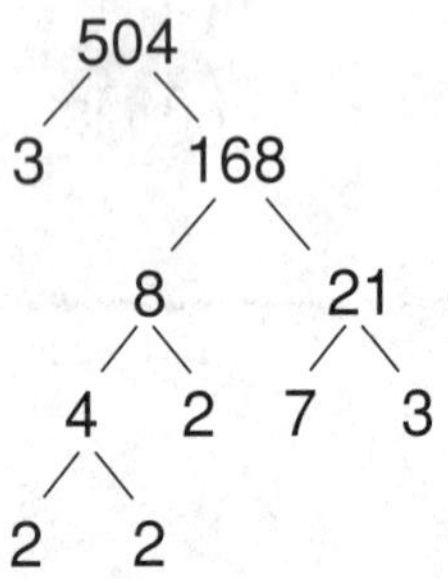

504

504

Prime factorization of 504: __________

3. In 1742, Christian Goldbach wrote his now famous Goldbach's Conjecture, which states that every even number greater than 2 can be represented as the sum of 2 primes. Although this conjecture is still an open question, it can be supported. Choose 3 even numbers greater than 2 to help support Goldbach's Conjecture.

____

4. Goldbach also made a conjecture that every odd number is the sum of 3 primes. Choose 3 odd numbers to support this conjecture.

____

Name ______________________ Date __________ Class __________

LESSON 2-6

Problem Solving

Prime Factorization

Write the correct answer.

1. The width of a swimming pool (in feet) is a prime number greater than 10. The width and length of the pool are factors of 408. What are the dimensions of the pool?

____

2. The area of the dining room at Thomas Jefferson's home in Monticello is about 342 square feet. If the approximate length of one side is a prime number less than 25, what are the approximate dimensions of the room?

____

3. A university has a lounge that can be converted into a meeting hall for 250 people. If the hall is filled and everyone is in equal groups, what are the different ways the people can be grouped so that there are no more than 10 groups?

____

____

4. You want to read a mystery that has 435 pages. If you read the same number of pages per day and the number is a prime number greater than 20, how many pages per day will you read?

____

____

Choose the letter of the best answer.

5. There are 228 seventh graders. Each seventh-grade homeroom starts the year with the same number of students and has at least 15 students. What is the least number of rooms that are needed?

A 17

B 19

C 12

D 21

6. Solve this riddle: I am a number whose prime factors are all the prime numbers between 6 and 15. No factor is repeated. What number am I?

F 9,009

G 91

H 1,001

J 6,006

7. What is the prime factorization of 1,485?

A 3 • 3 • 3 • 5 • 11

B 3 • 3 • 5 • 5 • 11

C 3 • 5 • 9 • 11

D 5 • 11 • 27

8. Solve this riddle: I am a prime factor of 39 and 65. What number am I?

F 3

G 5

H 11

J 13

Name ______________________ Date __________ Class __________

LESSON 2-6

Reading Strategies

Compare and Contrast

Whole numbers are either **prime numbers** or **composite numbers.** Compare the differences between them.

Whole Numbers

Prime Number
- Has only two factors: itself and 1.

Examples

$2 \rightarrow 2 \cdot 1 = 2$

$5 \rightarrow 1 \cdot 5 = 5$

Composite Number
- Has more than two factors.

Examples

$6 \rightarrow 1 \cdot 6 = 6$

$2 \cdot 3 = 6$

$9 \rightarrow 1 \cdot 9 = 9$

$3 \cdot 3 = 9$

A **composite number** can be written as the product of prime numbers. This is called **prime factorization.**
A **factor tree** helps you find all prime factors of a number.

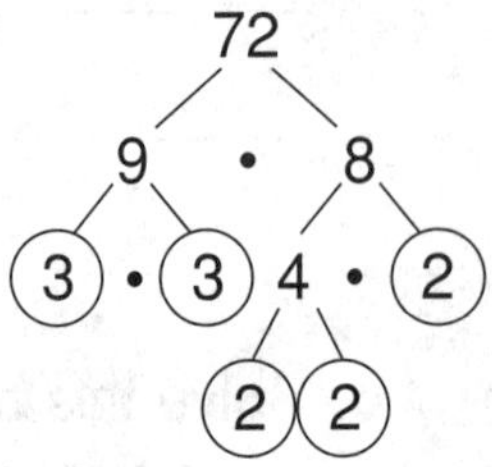

3 is a prime number. → Circle both 3's.
2 is a prime number. → Circle both 2's.

← 9 and 8 are not prime numbers. Continue to factor.
← 2 is a prime number. Circle 2. 4 is not a prime number. Continue to factor.

The prime factorization of 72 is $3 \cdot 3 \cdot 2 \cdot 2 \cdot 2$, or $3^2 \cdot 2^3$.

Answer each question.

1. What is the difference between a prime number and a composite number?

____

____

2. Find the prime factorization of 48. ______________________

3. Find the prime factorization of 56. ______________________

Name ______________________________ Date ____________ Class ____________

LESSON 2-6

Puzzles, Twisters & Teasers

Can You Find All the Words?

Solve the word search, then find the word from the list that best completes the riddle.

step	unique	composite	factor	factorization
exponent	tree	prime	diagram	Hajratwala

P	R	I	M	E	F	U	N	I	Q	U	E	P
C	W	E	S	A	H	D	C	B	N	U	O	F
T	Y	P	L	M	A	S	X	D	S	R	W	K
E	R	T	Y	U	J	C	F	T	S	Y	C	B
F	A	C	T	O	R	I	Z	A	T	I	O	N
A	S	Z	R	A	A	Q	W	R	E	I	M	F
C	G	B	E	X	T	D	S	A	P	B	P	M
T	N	M	E	A	W	N	J	I	L	K	O	C
O	W	E	R	T	A	T	H	N	B	V	S	I
R	Y	U	I	O	L	C	V	B	N	M	I	L
A	S	E	D	I	A	G	R	A	M	F	T	W
W	S	C	E	X	P	O	N	E	N	T	E	J

What was the mathematician's favorite food?

_____ _____ _____ _____ _____ rib

Name ______________________________ Date ____________ Class ____________

LESSON 2-7

Practice A

Greatest Common Factor

Draw a line to the greatest common factor (GCF). For Exercises 1–10, choose from A–K; for Exercises 11–20, choose from M–W.

1. 21, 35	**A.** 3	**11.** 18, 48	**M.** 16
2. 12, 32	**B.** 7	**12.** 10, 25	**N.** 60
3. 27, 66	**C.** 5	**13.** 120, 180	**P.** 6
4. 6, 12, 30	**D.** 4	**14.** 16, 48, 80	**Q.** 12
5. 20, 45, 75	**E.** 9	**15.** 6, 16, 30	**R.** 5
6. 27, 36, 45	**F.** 6	**16.** 60, 84, 120	**S.** 3
7. 25, 75, 100	**G.** 19	**17.** 42, 63, 84	**T.** 9
8. 38, 57	**H.** 8	**18.** 64, 96	**U.** 2
9. 32, 40, 120	**J.** 25	**19.** 18, 99, 135	**V.** 21
10. 36, 48, 60	**K.** 12	**20.** 24, 45, 90	**W.** 32

21. A jewelry designer is making identical bracelets. She has 21 blue beads and 14 red beads. What is the greatest number of bracelets she can make using all of the beads?

____

22. A florist is preparing identical flower baskets for a banquet. He has 72 tulips and 60 roses. What is the greatest number of flower baskets he can prepare using all of the tulips and roses?

____

Name ______________________ Date __________ Class __________

LESSON **2-7**

Practice B
Greatest Common Factor

Find the greatest common factor (GCF).

1. 12, 15 ________

2. 22, 33 ________

3. 63, 45 ________

4. 15, 50 ________

5. 18, 81 ________

6. 18, 48 ________

7. 20, 24 ________

8. 14, 42, 49 ________

9. 3, 6, 9 ________

10. 16, 24, 30 ________

11. 16, 40, 88 ________

12. 42, 70 ________

13. 25, 125, 200 ________

14. 26, 39, 52 ________

15. 36, 100 ________

16. 35, 77 ________

17. 56, 84 ________

18. 14, 49, 56, 84 ________

19. 30, 75, 60, 90 ________

20. 12, 38, 40, 94 ________

21. 48, 66, 96, 102 ________

22. Volunteers are preparing identical backpacks for refugees. There are 32 maps and 24 dictionaries to use for the backpacks. What is the greatest number of backpacks they can prepare using all of the maps and dictionaries?

____

23. Alyssa is preparing identical fruit baskets. There are 36 oranges and 60 apples to use for the baskets. What is the greatest number of fruit baskets she can prepare using all of the oranges and apples?

____

Name ______________________ Date __________ Class __________

LESSON 2-7

Practice C

Greatest Common Factor

Find the greatest common factor (GCF).

1. 45, 75 ____________

2. 24, 56, 72 ____________

3. 14, 18, 22 ____________

4. 60, 80, 120 ____________

5. 32, 64, 80 ____________

6. 32, 40, 120 ____________

7. 55, 115 ____________

8. 64, 96 ____________

9. 14, 35, 56, 77 ____________

10. Which pair of numbers has a GCF that is a prime number, 91 and 117 or 84 and 156?

____

11. Which pair of numbers has a GCF that is a prime number, 54 and 99 or 80 and 114?

____

12. Which pair of numbers has a GCF that is a prime number, 70 and 112 or 68 and 187?

____

13. Carvin is preparing identical breakfast gift sets. There are 18 boxes of pancake mix and 27 jars of jam. What is the greatest number of gift sets he can prepare using all of the pancake mix and jam?

____

14. Amelia is decorating identical frames. She has 75 red feathers and 30 blue tiles. What is the greatest number of frames she can decorate using all of the feathers and tiles?

____

15. A jeweler is making identical necklaces. He has 24 gold beads and 144 colored glass beads. What is the greatest number of necklaces he can make using all the gold beads and glass beads?

____

Name ______________________ Date __________ Class __________

LESSON 2-7

Reteach

Greatest Common Factor

The **greatest common factor** (GCF) of a group of numbers is the greatest number that can divide each of the numbers without having a remainder.

Use a list to find the GCF of 45, 60, and 75.
List all the factors of each number.

45:	1,		3,		5,		9,			15,			45
60:	1,	2,	3,	4,	5,	6,		10,	12,	15,	20,	30,	60
75:	1,		3,		5,					15,	25,		75

The common factors are 1, 3, 5, and 15.
The GCF is 15.

List the factors of each number. Find the greatest common factor (GCF).

1. 9: 1, ___, ___

24: 1, 2, ___, 4, ___, ___, ___, 24

GCF = ___

2. 18: 1, 2, 3, ___, ___, ___

54: 1, 2, 3, ___, ___, ___, ___, 54

GCF = ___

3. 45: 1, ___, ___, ___, ___, 45

55: 1, ___, ___, 55

GCF = ___

4. 12: 1, ___, ___, ___, ___, 12

20: 1, ___, ___, ___, ___, 20

GCF = ___

5. 6: 1, ___, ___, ___

12: 1, 2, ___, ___, ___, 12

24: 1, 2, ___, ___, ___, ___, 12, 24

GCF = ___

6. 6: 1, ___, ___, ___

18: 1, ___, ___, ___, ___, 18

27: 1, ___, ___, 27

GCF = ___

Find the greatest common factor (GCF).

7. 27, 45 ___________

8. 14, 18 ___________

9. 12, 30 ___________

10. 10, 45, 70 ___________

Name ______________________ Date ____________ Class ____________

LESSON **2-7**

Challenge

The School Trip Factor

There are 24 students in the seventh grade and 36 students in the eighth grade. Solve each problem below about their school trip. Then use your solutions to complete the permission slip each student must have signed in order to go on the trip.

1. The students will visit the zoo, the museum, or the planetarium. Each class will buy its own tickets. Both classes must agree upon the trip that will cost the least. Use the table at right to decide which trip the students will choose.

Trip	Single Tickets	Group Rate	Number of Tickets per Group
Zoo	$8	$24	6
Museum	$5	$12	4
Planetarium	$4	$30	12

2. The principal will provide transportation for both classes. The school PTA will pay the cost but wants the most cost-efficient method of transportation. Use the table at right to decide which method of transportation the PTA will choose.

Type of Transportation	Number of Passengers	Cost
Mini-bus	12	$35
School Bus	70	$185
Mini-van	10	$30

3. The class student council will provide lunch for each class. Each class must agree on the lowest price and then buy at least one lunch pack per student. Use the table at right to decide which lunch deal the students will choose.

Lunch	Number of Lunches per Pack	Cost per Pack
Sandwich and Juice Packs	6	$15
Yogurt and Drink Packs	8	$20
Soup and Sandwich Packs	4	$12

To the Seventh and Eighth-grade Parents:

On Friday, your child will be taking a school trip to the ___________.
The school will pay $__ for the seventh-graders tickets and $ __ for the eigth-graders tickets. The PTA has graciously offered to pay $___ for _ [number] __________ [vehicles] to transport the students. Each class s student council will pay for its class s lunches. Both classes will buy __ [number] __________________ [type] packs at a total cost of $ __ for the seventh graders and $__ for the eighth graders.

_______________________ Parent s signature for student s permission

Name ______________________ Date __________ Class __________

LESSON 2-7

Problem Solving

Greatest Common Factor

Write the correct answer.

1. Fabric is sold in stores from bolts that are 45 or 60 inches wide. What is the width of the widest strips of fabric you can cut from either bolt without wasting any of the fabric if each strip has the same width?

____

2. The parents are making sandwiches for the class picnic. They have 72 turkey slices, 48 cheese slices, and 96 tomato slices. What is the greatest number of sandwiches they can make if each sandwich has the same filling?

____

3. Two bicycle enthusiasts are leaving Cincinnati at the same time. One is biking 840 miles to Boston. The other is biking 440 miles to Atlanta. What is the greatest number of miles a day each can bike if they want to cover equal distances each day?

____

4. A fruit salad made on a TV cooking program requires chunks of cantaloupe and honeydew. What is the greatest number of servings you can make using all of the fruit if you have 30 chunks of cantaloupe and 42 chunks of honeydew?

____

Choose the letter for the best answer.

5. Ari is making patriotic pins. He has 105 red ribbons, 147 white ribbons, and 189 blue ribbons. What is the greatest number of identical pins he can make if he uses all his ribbons?

A 52 pins
B 19 pins
C 93 pins
D 21 pins

6. Cheryce is making fruit baskets. She has 60 bananas, 72 pears, 96 apples, and 108 oranges. What is the greatest number of equal baskets she can make with the fruit?

F 3 baskets
G 6 baskets
H 12 baskets
J 24 baskets

7. There are 100 senators and 435 representatives in the United States Congress. How many identical groups could be formed from all the senators and representatives?

A 1 group
B 5 groups
C 10 groups
D 15 groups

8. There are 14 baseball teams in the American League and 16 teams in the National League. There are 30 National Hockey League teams. If equal groups of teams are formed, how many hockey teams will be in each group?

F 3 hockey teams
G 8 hockey teams
H 7 hockey teams
J 15 hockey teams

Name ______________________ Date __________ Class __________

LESSON 2-7

Reading Strategies

Follow a Procedure

The **greatest common factor** of two numbers is the largest factor the numbers share or have in common. To find the greatest common factor for 36 and 45, follow these steps.

Step 1: Make a list of the factors for each number. →

36	**45**
1 • 36	1 • 45
2 • 18	3 • 15
3 • 12	5 • 9
4 • 9	
6 • 6	

Step 2: Put the factors in order from smallest to largest.

36: **1,** 2, **3,** 4, 6, **9,** 12, 18, 36
45: **1, 3,** 5, **9,** 15, 45

Step 3: Find the common factors or both numbers. →
1, 3, and **9**

Step 4: Find the greatest common factor for both numbers.
9

Answer each question.

1. What are the factors for 18? ______________

2. Put them in order from least to greatest. ______________

3. What are the factors for 21? ______________

4. Put them in order from least to greatest. ______________

5. Compare your answers for Exercise 2 and Exercise 4. What is the greatest common factor? ______________

6. Make a list of all the factors for 48.

48

7. Make a list of all the factors for 60.

60

8. Compare the lists. What factors do both numbers have in common? ______________

9. What is the greatest common factor for 48 and 60? ______________

Name ______________________ Date __________ Class __________

LESSON **2-7**

Puzzles, Twisters & Teasers

That's the Greatest!

Find the greatest common factor for each set of numbers. Then use the decoder to answer the riddle below.

1. 24, 36, 48, 60 __________ L

2. 30, 45, 60, 75 __________ E

3. 20, 40, 50, 120 __________ K

4. 4, 6, 10, 22 __________ N

5. 6, 15, 33, 48 __________ A

6. 14, 21, 35, 70 __________ R

7. 2, 3, 4, 5, 7 __________ O

8. 18, 45, 63, 81 __________ C

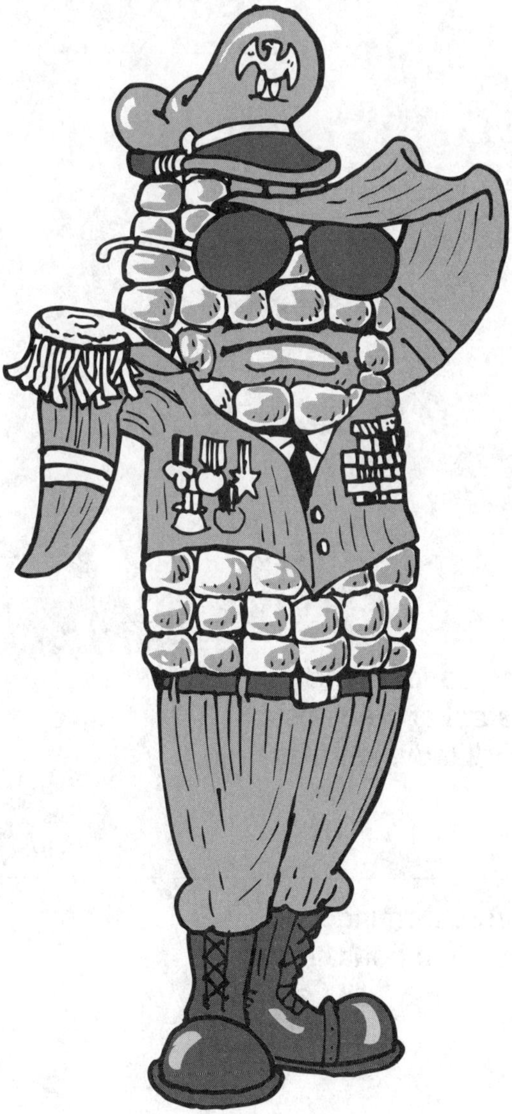

What do you call a piece of corn in the army?

___   ___ ___ ___ ___
3   9 1 7 2

___ ___ ___ ___ ___ ___
10 15 7 2 15 12

Name ______________________ Date __________ Class __________

LESSON **2-8**

Practice A

Least Common Multiple

Find the correct least common multiple (LCM) for each set of numbers below. Match each LCM with a number in the box.

21		**60**		**33**		**20**		**35**		**75**	
	42		**20**		**40**		**78**		**99**		**36**

1. 3, 7 ________

2. 4, 5 ________

3. 10, 12 ________

4. 15, 25 ________

5. 3, 11 ________

6. 9, 12, 18 ________

7. 6, 7, 21 ________

8. 5, 8, 20 ________

9. 2, 26, 39 ________

Find the LCM.

10. 10, 18, 45 ________

11. 5, 6, 20 ________

12. 10, 15, 30 ________

13. 2, 4, 7, 8 ________

14. 3, 9, 12 ________

15. 2, 5, 7, 10 ________

16. 5, 9, 18 ________

17. 4, 10, 12, 15 ________

18. 4, 9, 12, 18 ________

19. 8, 16, 32 ________

20. 5, 15, 20 ________

21. 2, 6, 18, 21 ________

22. Carrie and Lance are playing the drums for the band. In the song, Carrie beats the bass drum every 3 beats. Lance beats the snare drum every 5 beats. On what beat will they first play their drums together?

____

23. Domingo is helping his father paint some boards. Domingo paints a board every 12 minutes. His father paints a board every 8 minutes. If they start together, how long before they will finish painting a board at the same time?

____

Name ______________________ Date __________ Class __________

LESSON 2-8

Practice B

Least Common Multiple

Find the least common multiple (LCM).

1. 8, 10 ________

2. 10, 15 ________

3. 6, 9 ________

4. 12, 16 ________

5. 18, 30 ________

6. 5, 11 ________

7. 15, 45 ________

8. 7, 28 ________

9. 4, 14 ________

10. 3, 10, 12 ________

11. 9, 36, 60 ________

12. 5, 15 ________

13. 7, 14, 49 ________

14. 8, 12, 24, 96 ________

15. 5, 25, 30 ________

16. 5, 9, 18 ________

17. 4, 10, 12, 15 ________

18. 4, 9, 12, 18 ________

19. 4, 12, 24, 36 ________

20. 24, 30, 48, 60 ________

21. 5, 9, 15, 18 ________

22. Jasmine is helping her father plant trees to create a border around the back yard. Jasmine plants a tree every 25 minutes, and her father plants a tree every 15 minutes. If they started together, how long before they would finish planting a tree at the same time?

____

23. Two dancers are rehearsing in a studio. One dancer's routine lasts 12 minutes. The other dancer's routine lasts 15 minutes. If they start together and take no breaks between their routines, how long before they start together again?

____

24. Evan and Renzo are swimming laps in the pool. It takes Evan 8 minutes to complete 1 lap and Renzo 6 minutes to complete 1 lap. They start together at the tops of their lanes. In how many minutes will they be together again at the tops of their lanes?

____

Name ______________________ Date __________ Class __________

LESSON 2-8

Practice C

Least Common Multiple

Find the least common multiple (LCM).

1. 9, 12 ________

2. 18, 60 ________

3. 9, 15 ________

4. 20, 100 ________

5. 7, 11 ________

6. 8, 12, 16 ________

7. 3, 7, 21 ________

8. 40, 100 ________

9. 5, 9, 18 ________

10. 2, 26, 39 ________

11. 5, 8, 12 ________

12. 20, 40, 50 ________

13. 8, 9, 12 ________

14. 4, 6, 9, 12 ________

15. 3, 9, 12, 72 ________

16. 4, 10, 12 ________

17. 5, 20, 25, 50 ________

18. 4, 7, 8, 28 ________

19. 15, 20, 30 ________

20. 2, 6, 9, 21 ________

21. 10, 32, 64, 80 ________

22. Josh and Eva are packing first-aid supplies. It takes Josh 12 minutes to fill a box with bandages. It takes Eva 8 minutes to fill a box with antibiotics. If they start packing boxes at the same time, how long will it be before they start filling a new box at the same time?

____

23. Sadie is helping her mother stain some shelves. Sadie stains a shelf every 25 minutes. Her mother stains a shelf every 15 minutes. If they start together, how long will it be before they finish staining a shelf at the same time?

____

24. It takes Amanda 14 minutes to climb up the hill and sled down. It takes Alvin 12 minutes to climb up the hill and sled down. If they start up the hill at the same time, how long will it be before they meet again at the starting point?

____

Name ______________________ Date __________ Class __________

LESSON **2-8**

Reteach

Least Common Multiple

The **least common multiple** (LCM) of two numbers is the least multiple that the two numbers have in common.

- To list the multiples of a number, multiply the number by 1, 2, 3, and so on.

	8 • 1	8 • 2	8 • 3	8 • 4	8 • 5
Multiples of 8	8	16	24	32	40

	6 • 1	6 • 2	6 • 3	6 • 4	6 • 5
Multiples of 6	6	12	18	24	30

- To find the least common multiple of 6 and 8, list several multiples of each number.

Step 1: List multiples of the greater number.

Multiples of 8: 8 16 **24** 32 40

Step 2: List the multiples of the lesser number until you find a multiple that is common to each list.

Multiples of 6: 6 12 18 **24**

Since 24 is the first number common to both lists, it is the least common multiple. So the LCM of 6 and 8 is 24.

Find the least common multiple (LCM).

1. 4, 9

Multiples of 9: 9, 18, 27, 36, 45, 54

Multiples of 4: 4, 8, 12, 16, 20, 24, 28, 32, 36

LCM of 4 and 9: ______

2. 6, 7

Multiples of 7: 7, 14, ______, ______, ______, ______, ______, ______

Multiples of 6: 6, 12, ______, ______, ______, ______, ______

LCM of 6 and 7: ______

3. 15, 20

Multiples of 20: ______, ______, ______, ______, ______

Multiples of 15: ______, ______, ______, ______

LCM of 15 and 20: ______

4. 12, 18 ______

5. 16, 20 ______

6. 4, 6, 9 ______

Name ______________________________ Date ____________ Class ____________

LESSON 2-8

Challenge

Multiple Solutions

Write the two numbers from each box that have the lowest least common multiple. Then write the LCM.

1. | 5 12 / 6 8 | → Numbers: __________ LCM: ______

2. | 2 7 / 9 10 | → Numbers: __________ LCM: ______

3. | 2 10 / 5 3 | → Numbers: __________ LCM: ______

4. | 7 2 / 10 12 | → Numbers: __________ LCM: ______

5. | 6 4 / 7 5 | → Numbers: __________ LCM: ______

6. | 9 10 / 15 45 | → Numbers: __________ LCM: ______

7. | 8 10 / 6 5 | → Numbers: __________ LCM: ______

8. | 24 35 / 32 14 | → Numbers: __________ LCM: ______

9. | 15 10 / 7 8 | → Numbers: __________ LCM: ______

10. | 3 10 / 12 7 | → Numbers: __________ LCM: ______

Use the table at the bottom of the page to solve the riddle. In the middle row, write the LCM of each exercise above the exercise number. Then choose the letter from the Answer Key that matches the LCM and write it in the appropriate box.

Answer Key

C	A	T	E	H
70	6	30	12	10

11. What animal copies from the others on a math test?

9	4	5		8	2	1	5	6	3	7

Name ______________________ Date __________ Class __________

LESSON 2-8

Problem Solving

Least Common Multiple

Write the correct answer.

1. Earth revolves around the sun every year. Jupiter revolves around the sun every 12 years. If Earth and Jupiter passed the same point of the sun sometime in 2002, when will they pass that point together again?

____

2. House representatives are elected every 2 years. The President of the United States is elected every 4 years. Both will be elected in 2008. When is the next year after 2008 both will be elected?

____

3. A cat runs a mile every 2 minutes. A squirrel runs a mile every 5 minutes. A cat and a squirrel start together running around a 1-mile track. How long will it be before they meet at the starting point?

____

4. A car manual recommends changing the oil every 5,000 miles and inspecting the engine coolant system every 15,000 miles. At how many miles will both be done together for the first time? for the second time?

____

Choose the letter for the best answer.

5. Mr. Walters receives a dividend every 5 months and a royalty payment every 6 months. He received both in January 2006. When is the next time he would receive both payments in the same month?

   A January 2007
   B April 2008
   C July 2006
   D July 2008

6. Rag Rite Cloth Store always rounds amounts less than whole yards up to the next yard for ribbon purchases. The ribbon that you want to buy comes in rolls of 8 feet. How many rolls should you buy to get the best buy?

   F 1 roll
   G 2 rolls
   H 3 rolls
   J 4 rolls

7. The sanitation department picks up recyclable plastics every 3 days. The paper recycling center picks up papers every 4 days. They both picked up on May 4. When will they next pick up on the same day?

   A May 7
   B May 12
   C May 11
   D May 16

8. Hal and Jess both volunteer at the local nursing home. Hal volunteers every 6 days, and Jess volunteers every 8 days. They were both there on Monday. In how many days will they both volunteer together again?

   F in 14 days
   G in 24 days
   H in 48 days
   J in 2 days

Name ______________________ Date __________ Class __________

LESSON **2-8**

Reading Strategies

Focus on Vocabulary

A **multiple** of a number is the product of that number and another factor.

Multiples of 5 are the numbers you get when you multiply by **5.**

$1 \cdot 5 = \mathbf{5}$ $2 \cdot 5 = \mathbf{10}$ $3 \cdot 5 = \mathbf{15}$ $4 \cdot 5 = \mathbf{20}$

The multiples of 5 listed above are 5, 10, 15, 20.

You can continue to count by 5's to list more multiples of 5: 25, 30, 35, and so on.

Common means something that is shared. A common multiple of two numbers is a multiple that the two numbers share, or have in common.

Multiples of 5 → 5, 10, 15, **20,** 25, 30, 35, **40,** 45, 50, 55, **60**

Multiples of 4 → 4, 8, 12, 16, **20,** 24, 28, 32, 36, **40,** 44, 48, 52, 56, **60**

In the example above, you can see the multiples that are common to both 4 and 5. They are 20, 40, and 60.

The **least common multiple (LCM)** of two numbers is the smallest of the common multiples. 20 is the smallest, or least common, multiple of 4 and 5.

Answer each question.

1. What is a multiple?

____

____

2. What does the word *common* mean?

____

3. What is a common multiple?

____

4. What is the least common multiple?

____

5. Write the first 9 multiples of 2. ______________________

6. Write the first 9 multiples of 3. ______________________

7. Write the common multiples for both 2 and 3. ______________________

8. What is the least common multiple for 2 and 3? ______________________

Name ______________________ Date ____________ Class ____________

LESSON 2-8

Puzzles, Twisters & Teasers

A Common Cause!

Find the least common multiple for each set of numbers. Spell out the answers to solve the crossword puzzle.

Across

1. 3, 5

4. 9, 12

6. 10, 4

7. 2, 5

9. 6, 14

Down

1. 3, 9, 15

2. 4, 6, 24

3. 6, 9

5. 4, 7

6. 3, 7

8. 15, 20

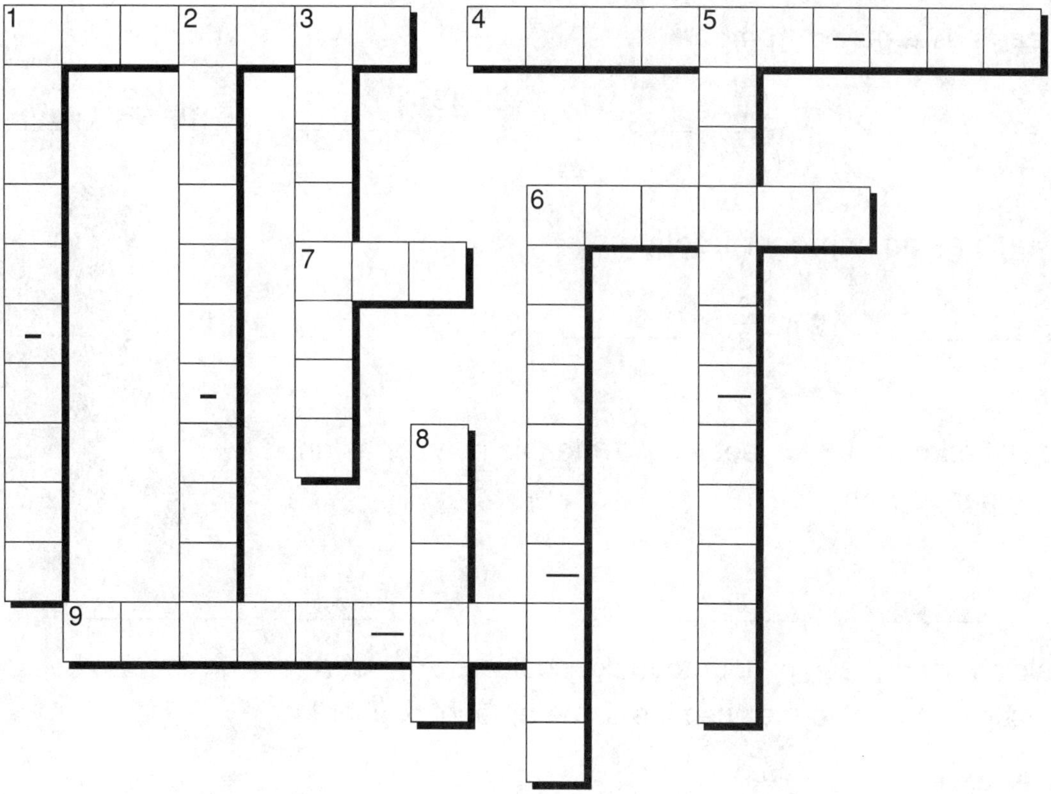

Name ______________________ Date __________ Class __________

LESSON 2-9

Practice A

Equivalent Fractions and Mixed Numbers

Find a fraction equivalent to each fraction.

1. $\frac{1}{4}$ __________ **2.** $\frac{3}{5}$ __________ **3.** $\frac{4}{9}$ __________ **4.** $\frac{1}{8}$ __________

5. $\frac{1}{6}$ __________ **6.** $\frac{20}{40}$ __________ **7.** $\frac{6}{15}$ __________ **8.** $\frac{35}{50}$ __________

Write the fractions with a common denominator. Then determine if they are equivalent.

9. $\frac{2}{3}$ and $\frac{6}{9}$ **10.** $\frac{7}{10}$ and $\frac{8}{15}$ **11.** $\frac{3}{6}$ and $\frac{4}{8}$

__________ __________ __________

12. $\frac{16}{8}$ and $\frac{10}{5}$ **13.** $\frac{9}{6}$ and $\frac{6}{4}$ **14.** $\frac{8}{12}$ and $\frac{15}{24}$

__________ __________ __________

Write each as a mixed number.

15. $\frac{5}{4}$ __________ **16.** $\frac{9}{2}$ __________ **17.** $\frac{13}{5}$ __________ **18.** $\frac{12}{7}$ __________

Write each as an improper fraction.

19. $3\frac{2}{3}$ __________ **20.** $2\frac{3}{8}$ __________ **21.** $1\frac{3}{5}$ __________ **22.** $2\frac{5}{6}$ __________

23. Lou's bedroom is $11\frac{1}{2}$ feet long. Write this number as an improper fraction.

____

24. Kaleigh used $\frac{2}{3}$ cup of flour to make cookies. Eric used $\frac{9}{12}$ cup of flour. Did they use the same amount of flour?

____

Name ______________________ Date ____________ Class ____________

LESSON 2-9

Practice B

Equivalent Fractions and Mixed Numbers

Find a fraction equivalent to the given fraction.

1. $\frac{2}{9}$ ________ **2.** $\frac{8}{15}$ ________ **3.** $\frac{7}{8}$ ________

4. $\frac{16}{24}$ ________ **5.** $\frac{12}{20}$ ________ **6.** $\frac{9}{12}$ ________

Write the fractions with a common denominator.
Then determine if they are equivalent.

7. $\frac{8}{10}$ and $\frac{12}{15}$ **8.** $\frac{6}{8}$ and $\frac{8}{12}$ **9.** $\frac{3}{9}$ and $\frac{4}{8}$

________ ________ ________

10. $\frac{7}{4}$ and $\frac{9}{5}$ **11.** $\frac{15}{12}$ and $\frac{20}{16}$ **12.** $\frac{15}{9}$ and $\frac{30}{18}$

________ ________ ________

Write each as a mixed number.

13. $\frac{21}{8}$ ________ **14.** $\frac{37}{4}$ ________ **15.** $\frac{16}{5}$ ________ **16.** $\frac{49}{9}$ ________

Write each as an improper fraction.

17. $8\frac{2}{3}$ ________ **18.** $1\frac{7}{12}$ ________ **19.** $25\frac{3}{4}$ ________ **20.** $7\frac{5}{6}$ ________

21. Maria's desk is $33\frac{3}{4}$ inches long. Write this number as an improper fraction.

____

22. Leon walked $\frac{5}{8}$ mile. Liz walked $\frac{10}{16}$ mile. Did they walk the same distance?

____

Name ______________________ Date ____________ Class ____________

LESSON 2-9

Practice C

Equivalent Fractions and Mixed Numbers

Find two fractions equivalent to each number.

1. $\frac{7}{12}$ ________

2. $\frac{11}{15}$ ________

3. $\frac{14}{25}$ ________

4. $\frac{18}{20}$ ________

5. 5 ________

6. $1\frac{1}{8}$ ________

Write the fractions with a common denominator. Then determine if they are equivalent.

7. $\frac{10}{15}$ and $\frac{12}{20}$

____

8. $\frac{10}{18}$ and $\frac{9}{12}$

____

9. $\frac{18}{24}$ and $\frac{12}{16}$

____

10. $\frac{7}{8}$ and $\frac{9}{10}$

____

11. $\frac{32}{6}$ and $\frac{48}{9}$

____

12. $\frac{38}{15}$ and $\frac{25}{10}$

____

Write each as a mixed number.

13. $\frac{43}{7}$ ________

14. $\frac{27}{5}$ ________

15. $\frac{55}{8}$ ________

16. $\frac{80}{11}$ ________

Write each as an improper fraction.

17. $25\frac{2}{5}$ ________

18. $11\frac{7}{9}$ ________

19. $4\frac{11}{12}$ ________

20. $5\frac{5}{16}$ ________

21. Write an improper fraction and a mixed number to represent 139 inches in feet.

____

22. A baker used $52\frac{1}{2}$ cups of flour to make 15 loaves of bread. How many quarter cups of flour were used?

____

Name ______________________ Date __________ Class __________

LESSON 2-9

Reteach

Equivalent Fractions and Mixed Numbers

Equivalent fractions name the same number.

- You can use fraction strips to find equivalent fractions.

Find two fractions equivalent to $\frac{6}{8}$.

$\frac{1}{8}$	$\frac{1}{8}$	$\frac{1}{8}$	$\frac{1}{8}$	$\frac{1}{8}$	$\frac{1}{8}$

First, use fraction strips to model $\frac{6}{8}$.

Then find equivalent fractions by placing other fraction strips with the same length below the model.

$\frac{1}{8}$		$\frac{1}{8}$		$\frac{1}{8}$		$\frac{1}{8}$		$\frac{1}{8}$		$\frac{1}{8}$		$\frac{6}{8}$
$\frac{1}{4}$				$\frac{1}{4}$				$\frac{1}{4}$				$\frac{3}{4}$
$\frac{1}{16}$	$\frac{1}{16}$	$\frac{1}{16}$	$\frac{1}{16}$	$\frac{1}{16}$	$\frac{1}{16}$	$\frac{1}{16}$	$\frac{1}{16}$	$\frac{1}{16}$	$\frac{1}{16}$	$\frac{1}{16}$	$\frac{1}{16}$	$\frac{12}{16}$

$\frac{3}{4}$ and $\frac{12}{16}$ are equivalent to $\frac{6}{8}$.

Use fractions strips to find a fraction equivalent to each fraction.

1. $\frac{1}{2}$ ________ **2.** $\frac{1}{4}$ ________ **3.** $\frac{2}{5}$ ________ **4.** $\frac{2}{3}$ ________

- You can use cross products to determine if two fractions are equivalent.

$\frac{8}{10}$ $\frac{12}{15}$ $\quad 8 \cdot 5 \overset{?}{=} 10 \cdot 12$

$120 = 120$

$\frac{8}{10}$ and $\frac{12}{15}$ are equivalent. Their cross products are equal.

$\frac{9}{15}$ $\frac{2}{3}$ $\quad 9 \cdot 3 \overset{?}{=} 15 \cdot 2$

$27 \neq 30$

$\frac{9}{15}$ and $\frac{2}{3}$ are not equivalent. Their cross products are not equal.

Find the cross products. Determine if the two fractions are equivalent. Write *yes* or *no*.

5. $\frac{18}{6}$ and $\frac{10}{3}$ **6.** $\frac{6}{10}$ and $\frac{9}{15}$ **7.** $\frac{2}{3}$ and $\frac{9}{12}$

________ ________ ________

Name ________________________ Date __________ Class __________

LESSON 2-9

Reteach

Equivalent Fractions and Mixed Numbers (continued)

In an *improper fraction* the numerator is greater than the denominator, such as $\frac{11}{3}$.

A *mixed number* is made up of both a whole number part and a fraction part, such as $2\frac{5}{8}$.

- To write an improper fraction as a mixed number, you can write the improper fraction as a sum of fractions.

$$\frac{11}{3} = \underbrace{\frac{3}{3} + \frac{3}{3} + \frac{3}{3}}_{1 + 1 + 1} + \frac{2}{3} = \frac{2}{3}$$

- To write a mixed number as an improper fraction, you can write the whole number as a fraction with the same denominator as the fraction part.

$$2\frac{5}{8} = 2 + \frac{5}{8} = \frac{16}{8} + \frac{5}{8} = \frac{21}{8}$$

8. Write $\frac{13}{4}$ as a mixed number.

$$\frac{13}{4} = \underset{\uparrow \atop 1}{\frac{\ \ }{4}} + \underset{\uparrow \atop 1}{\frac{\ \ }{4}} + \underset{\uparrow \atop 1}{\frac{\ \ }{4}} + \frac{\ \ }{4} = \_\_\_\_\_$$

9. Write $3\frac{2}{5}$ as an improper fraction.

$$3\frac{2}{5} = \_\_\_\_\_ + \frac{\ \ }{\ \ } = \frac{\ \ }{5} + \frac{\ \ }{\ \ } = \frac{\ \ }{\ \ }$$

Write each improper fraction as a mixed number.

10. $\frac{23}{5}$ _____ **11.** $\frac{15}{8}$ _____ **12.** $\frac{54}{7}$ _____ **13.** $\frac{14}{3}$ _____

Write each mixed number as an improper fraction.

14. $5\frac{1}{2}$ _____ **15.** $3\frac{3}{8}$ _____ **16.** $2\frac{11}{15}$ _____ **17.** $8\frac{3}{4}$ _____

Name ______________________ Date ____________ Class ____________

LESSON **2-9**

Challenge

Target Practice

Find out where the dart lands on the dartboard. Cross out the fraction, improper fraction, or mixed number on the dartboard that is equivalent to each number below. The dart lands on the number that is left.

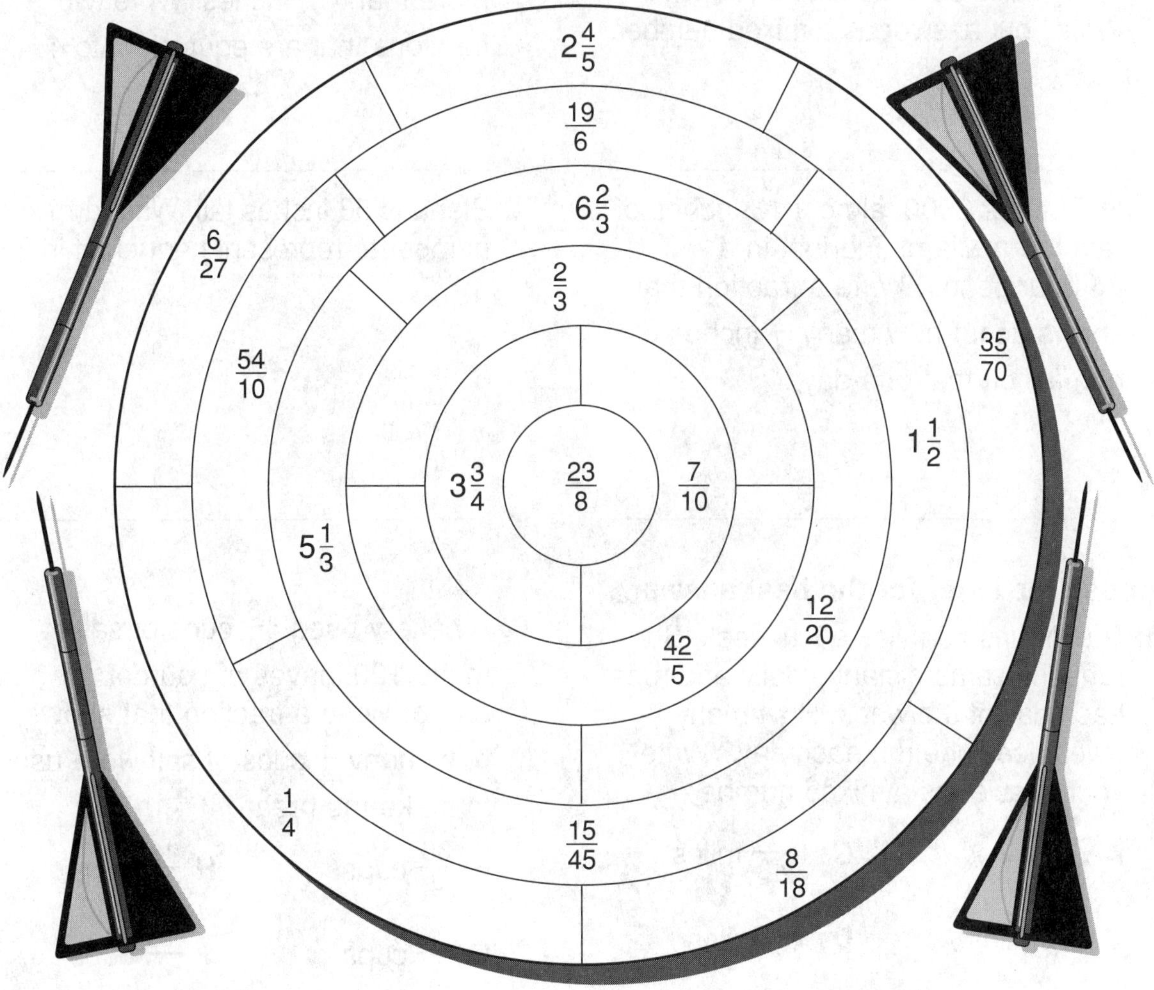

1. $\frac{32}{6}$ **2.** $\frac{14}{20}$ **3.** $\frac{12}{8}$ **4.** $\frac{3}{5}$

5. $2\frac{7}{8}$ **6.** $\frac{28}{10}$ **7.** $\frac{16}{36}$ **8.** $\frac{50}{75}$

9. $\frac{2}{9}$ **10.** $\frac{20}{3}$ **11.** $\frac{5}{15}$ **12.** $3\frac{1}{6}$

13. $\frac{3}{12}$ **14.** $5\frac{2}{5}$ **15.** $\frac{45}{12}$ **16.** $\frac{1}{2}$

17. The dart lands on ____________

Name ______________________ Date __________ Class __________

LESSON 2-9

Problem Solving

Equivalent Fractions and Mixed Numbers

Write the correct answer.

1. Bert bicycles 136 miles each week. If he bikes the same distance daily and bikes each day of the week, how many miles does he bike each day? Write your answer as a mixed number.

____

2. The Eiffel Tower in Paris, France, was designed so well that even in the highest winds the tower never sways more than $4\frac{1}{2}$ inches. Write two fractions that are equivalent to $4\frac{1}{2}$.

____

3. In October 2000, almost 13 inches of rain fell in Miami, Florida, in a 24-hour period. Write a fraction that shows about how many $\frac{1}{4}$ inches of rain fell on that one day.

____

4. Elena is 58 inches tall. Write a mixed number to represent her height in feet.

____

Choose the letter for the best answer.

5. Randi runs 48 miles each week. She runs the same distance daily and runs each day of the week. How many miles does she run each day? Write your answer as a mixed number.

A $6\frac{6}{7}$ miles **C** $6\frac{7}{48}$ miles

B $6\frac{5}{6}$ miles **D** $7\frac{1}{7}$ miles

6. A bakery used $4\frac{3}{4}$ cups of salt to make 120 loaves of sourdough bread. Write a fraction that shows how many $\frac{1}{4}$ cups of salt were used to make the bread.

F $\frac{30}{4}$ cups **H** $\frac{12}{4}$ cups

G $\frac{24}{4}$ cups **J** $\frac{19}{4}$ cups

7. One of the driest summers in the Northeast was in 1909, with $\frac{896}{100}$ inches of precipitation. Which mixed number is equivalent to $\frac{896}{100}$?

A $89\frac{6}{10}$ **C** $8\frac{24}{25}$

B $8\frac{96}{10}$ **D** $8\frac{4}{5}$

8. The Central Ohio Transit Authority has $957\frac{3}{5}$ miles of routes. Which fraction is equivalent to $957\frac{3}{5}$?

F $\frac{4,781}{5}$ **H** $\frac{4,783}{5}$

G $\frac{4,788}{5}$ **J** $\frac{9,573}{5}$

Name ______________________________ Date ____________ Class ____________

LESSON 2-9

Reading Strategies

Multiple Meaning Terms

If you multiply or divide the numerator and denominator of a fraction by the same number, you get another fraction that is **equivalent,** or equal.

$\frac{2}{3} \cdot \frac{2}{2} = \frac{4}{6}$

In the example above, both the numerator and denominator are multiplied by 2. Equal fractions are called **equivalent fractions.**

$\frac{2}{3}$ is equivalent to $\frac{4}{6}$.

Answer the following questions.

1. Is $\frac{1}{2}$ equivalent to $\frac{6}{12}$? ____________

2. How do you know?

____

____

3. Are $\frac{1}{5}$ and $\frac{2}{3}$ equivalent fractions? ____________

4. How do you know?

____

____

5. Is $\frac{7}{15}$ equivalent to $\frac{1}{7}$? ____________

6. How do you know?

____

____

7. Is $\frac{3}{9}$ equivalent to $\frac{1}{3}$? ____________

8. How do you know?

____

____

Name ______________________ Date __________ Class __________

LESSON 2-9

Puzzles, Twisters & Teasers

Are You in the Dark?

Fill in the boxes in the grids to show equivalent fractions. Then use the Answer Key to answer the riddle.

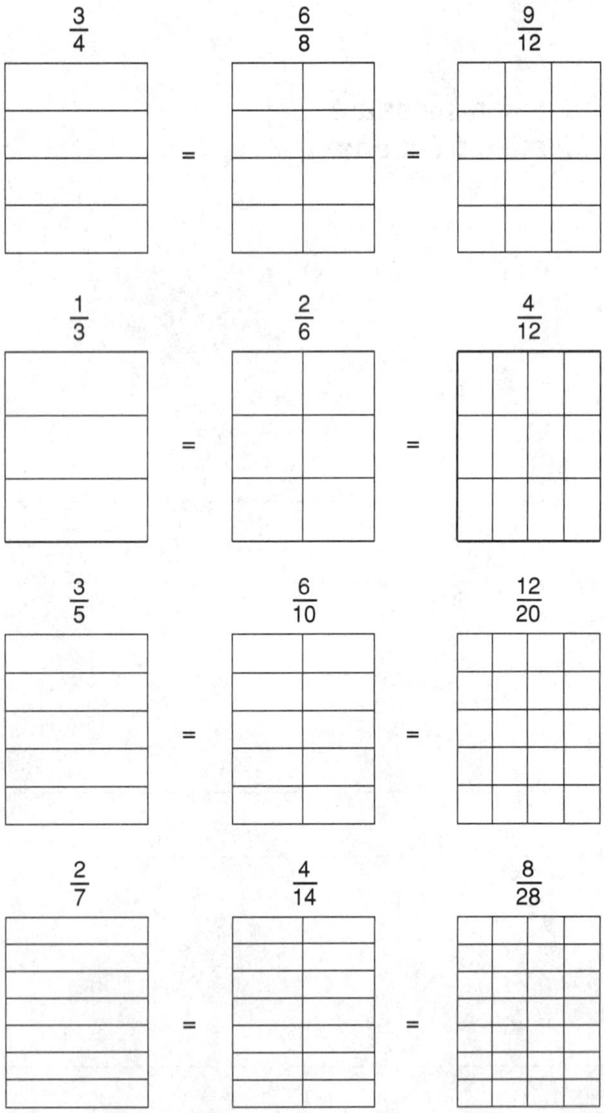

1I **2**T **3**S **4**A **6**E **8**R **12**D

Why did the tomato blush?

___ ___   ___ ___ **W** ___ **H** ___
1 2   3 4   2   6

___ ___ **L** ___ ___   ___ ___ ___ ___ ___ ___ **N G .**
3 4   4 12   12 8 6 3 3 1

Name ______________________ Date ____________ Class ____________

LESSON **2-10**

Practice A

Equivalent Fractions and Decimals

Write each fraction as a decimal. Round to the nearest hundredth, if necessary.

1. $\frac{7}{10}$ ________ **2.** $\frac{9}{20}$ ________ **3.** $\frac{3}{4}$ ________ **4.** $\frac{20}{25}$ ________

5. $\frac{3}{8}$ ________ **6.** $\frac{7}{5}$ ________ **7.** $\frac{21}{7}$ ________ **8.** $\frac{5}{3}$ ________

9. $\frac{3}{10}$ ________ **10.** $\frac{4}{5}$ ________ **11.** $\frac{1}{25}$ ________ **12.** $\frac{3}{20}$ ________

Write each decimal as a fraction or mixed number in simplest form.

13. 0.55 **14.** 0.03 **15.** −0.75

________ ________ ________

16. 2.1 **17.** 5.25 **18.** 9.33

________ ________ ________

19. 1.8 **20.** −1.74 **21.** 10.6

________ ________ ________

22. −7.08 **23.** 0.625 **24.** 0.001

________ ________ ________

Write each answer as a decimal rounded to the nearest thousandth.

25. Out of 45 times at bat, Raul got 19 hits. Find Raul's batting average.

____

26. On a test, Selena answered 26 out of 30 questions correctly. What portion of her answers was correct?

____

Name ______________________ Date __________ Class __________

LESSON **2-10**

Practice B

Equivalent Fractions and Decimals

Write each fraction as a decimal. Round to the nearest hundredth, if necessary.

1. $\frac{2}{10}$ ________ **2.** $\frac{19}{20}$ ________ **3.** $\frac{5}{8}$ ________ **4.** $\frac{11}{5}$ ________

5. $\frac{19}{6}$ ________ **6.** $\frac{17}{4}$ ________ **7.** $\frac{13}{12}$ ________ **8.** $\frac{30}{7}$ ________

9. $\frac{7}{4}$ ________ **10.** $\frac{9}{20}$ ________ **11.** $\frac{11}{10}$ ________ **12.** $\frac{2}{25}$ ________

Write each decimal as a fraction in simplest form.

13. 0.85 ________ **14.** 0.11 ________ **15.** −0.25 ________

16. 4.3 ________ **17.** 7.75 ________ **18.** 5.03 ________

19. −1.06 ________ **20.** 0.375 ________ **21.** −2.65 ________

22. −5.6 ________ **23.** 1.12 ________ **24.** 0.005 ________

Write each answer as a decimal rounded to the nearest thousandth.

25. In the 1998 Winter Olympics, a total of 205 medals were awarded. The United States won 13 medals. What portion of the medals did the United States win?

____

26. On a test, Hailey answered 64 out of 75 questions correctly. What portion of her answers was correct?

____

Name ______________________ Date __________ Class __________

LESSON 2-10

Practice C

Equivalent Fractions and Decimals

Write each fraction as a decimal. Round to the nearest hundredth, if necessary.

1. $\frac{5}{12}$ ________ **2.** $\frac{19}{4}$ ________ **3.** $\frac{25}{7}$ ________ **4.** $\frac{9}{8}$ ________

5. $\frac{17}{3}$ ________ **6.** $\frac{13}{11}$ ________ **7.** $\frac{15}{9}$ ________ **8.** $\frac{30}{13}$ ________

9. $\frac{9}{50}$ ________ **10.** $\frac{13}{25}$ ________ **11.** $\frac{23}{20}$ ________ **12.** $\frac{17}{4}$ ________

Write each decimal as a fraction in simplest form.

13. 0.25 ________ **14.** 0.13 ________ **15.** −0.65 ________

16. 4.8 ________ **17.** 1.875 ________ **18.** 9.06 ________

19. −1.008 ________ **20.** 0.625 ________ **21.** −2.05 ________

Determine if the numbers in each pair are equivalent.

22. $\frac{3}{7}$ and 0.41 ________ **23.** $\frac{13}{20}$ and 0.65 ________ **24.** $\frac{19}{25}$ and 0.78 ________

25. In the 1996 Summer Olympics, a total of 842 medals were awarded. The United States won 101 medals. What portion of the medals did the United States win? Write your answer as a decimal rounded to the nearest thousandth.

____

26. Pete Rose had 4,256 hits in 14,053 at bats. Write his batting average as a fraction and as a decimal rounded to the nearest thousandth.

____

Name ______________________ Date ___________ Class ___________

LESSON 2-10

Reteach

Equivalent Fractions and Decimals

To write a fraction as a decimal, divide the numerator of the fraction by the denominator of the fraction.

Write $\frac{3}{7}$ as a decimal.

- Divide 3 by 7.
- To round your answer to the nearest hundredth, add 3 zeros after the decimal point in the divisor.

$$\begin{array}{r} 0.428 \\ 7\overline{)3.000} \\ -\underline{2\,8}\downarrow \\ 20\downarrow \\ -\underline{14} \\ 60\downarrow \\ -\underline{56} \end{array}$$

0.428 rounded to the nearest hundredth is 0.43.

1. Write $\frac{2}{5}$ as a decimal. $\frac{2}{5}$ = ___________

$$\begin{array}{r} 5\overline{)2\,.\,0} \\ -\underline{\quad\quad} \\ \underline{\quad\quad} \end{array}$$

Write each fraction as a decimal. Round to the nearest hundredth, if necessary.

2. $\frac{3}{4}$ ________ **3.** $\frac{7}{8}$ ________ **4.** $\frac{3}{2}$ ________ **5.** $\frac{5}{3}$ ________

To write a decimal as a fraction:

Step 1: Use place value to read the decimal. Say the number aloud.

Step 2: Write a fraction for the number you just said.

Step 3: Simplify if necessary.

Write 0.005 as a fraction.

Read 0.005 as "five thousandths."

Write $\frac{5}{1000}$ for five thousandths.

Simplify: $\frac{5 \div 5}{1{,}000 \div 5} = \frac{1}{200}$

Write 1.6 as a fraction.

Read 1.6 as "one and six tenths."

Write $1\frac{6}{10}$ for one and six tenths.

Simplify: $1\frac{6 \div 2}{10 \div 2} = 1\frac{3}{5}$

Write each decimal as a fraction or mixed number in simplest form.

6. 0.8 ___________ **7.** 2.25 ___________ **8.** −0.02 ___________

Name ______________________ Date ____________ Class ____________

LESSON 2-10

Challenge

Decimal Patterns

Repeating decimals continue without end with a repeating pattern. Some decimals continue without end, but do not have a repeating pattern. These decimals are called *nonrepeating, nonterminating* decimals. Although they continue without end, these decimals may have a pattern.

Examples of nonrepeating, nonterminating decimals with patterns:

5.121121112...
0.881882883...

Write *r* if the decimal is a repeating decimal and *n* if the decimal is a nonrepeating, nonterminating decimal.

1. 4.562222... ____________

2. 6.1323323332... ____________

3. 8.355355... ____________

4. 0.230230023... ____________

5. 10.7727227... ____________

6. 4.121314... ____________

7. 0.42818181... ____________

8. 29.7053053... ____________

9. 1.221221222... ____________

10. 5.67167672... ____________

11. 75.320333... ____________

12. 4.121231234... ____________

Identify the pattern in each nonrepeating, nonterminating decimal. Use the pattern to extend the decimal several places.

13. 0.010203...

____

14. 8.636336333...

____

15. 25.121231234...

____

16. 9.989796...

____

17. Write an example of a repeating decimal.

____

18. Write an example of a nonrepeating, nonterminating decimal with a pattern.

____

Name ______________________ Date __________ Class __________

LESSON **2-10**

Problem Solving

Equivalent Fractions and Decimals

Write the correct answer.

1. On a test, Shane answered 37 out of 40 questions correctly. What portion of his answers was correct? Write your answer as a decimal rounded to the nearest thousandth.

____

2. Sammy Sosa hit 66 home runs in 1998. He had 643 at bats. Write his home run average as a decimal rounded to the nearest thousandth.

____

3. In February, Chicago receives an average of $1\frac{2}{5}$ inches of rain. Write a decimal to show the number of inches of rain.

____

4. On a test, Ellen answered 51 out of 64 questions correctly. What portion of her answers was correct? Write your answer as a decimal rounded to the nearest thousandth.

____

Choose the letter for the best answer.

Use the graph for 5–6.

World Energy Consumption 2003
(in quadrillions of British Thermal Units, BTU)

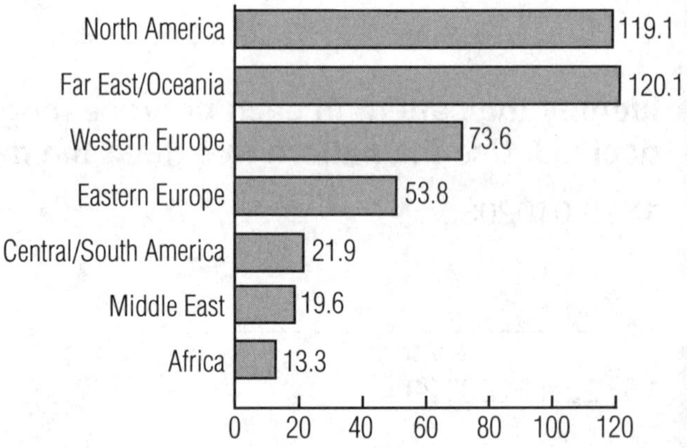

5. Which mixed number shows the energy, in quadrillions of BTU, consumed in Eastern Europe in 2003?

A 53 **C** $53\frac{1}{2}$

B $53\frac{1}{4}$ **D** $53\frac{4}{5}$

6. Which mixed number shows the energy, in quadrillions of BTU, consumed in the Middle East in 2003?

F $19\frac{1}{10}$ **H** $19\frac{3}{5}$

G $19\frac{2}{5}$ **J** $19\frac{3}{10}$

7. Jill sold 478 out of 520 tickets to the opening night of her theater performance. What portion of the tickets did she sell?

A 0.998 **C** 0.919

B 1.088 **D** 0.081

8. The high school sold 369 out of 460 tickets to the opening night of a concert series. What portion of the tickets was sold?

F 0.198 **H** 1.247

G 0.829 **J** 0.802

Name ______________________ Date __________ Class __________

LESSON **2-10**

Reading Strategies

Compare and Contrast

Compare what happens when fractions are changed to decimals.

$\frac{2}{5}$ • Read $\frac{2}{5}$ as "2 divided by 5." • Write ⟶ $2 \div 5$

Change a fraction to a decimal by dividing the numerator by the denominator.

$$\begin{array}{r} .4 \\ 5\overline{)2.0} \\ -2\,0 \end{array}$$

$\frac{2}{5} = 0.4$ The dividing ends, or terminates, with no remainder. 0.4 is called a terminating decimal.

1. Is there a remainder in the problem? How do you know?

____

____

2. What do we call a decimal that ends with no remainder?

____

$\frac{2}{6}$ • Read $\frac{2}{6}$ as "2 divided by 6." • Write ⟶ $2 \div 6$

$$\begin{array}{r} .333 \\ 6\overline{)2.000} \\ \underline{-1\,8} \\ 20 \\ \underline{-18} \\ 20 \\ \underline{-18} \\ 2 \end{array}$$

$\frac{2}{6} = 0.333\ldots$ or $0.\overline{3}$

⟵ Note how dividing continues in a pattern. The number 0.333 . . . is a repeating decimal. The bar over the 3 means 3 repeats.

Answer each question.

3. Compare the division of $\frac{2}{5}$ to the division of $\frac{2}{6}$. What is the difference?

____

____

4. What is the name for a decimal with a remainder that has a repeating pattern?

____

Name ______________________________ Date ____________ Class ____________

LESSON **2-10**

Puzzles, Twisters & Teasers

The Terminator!

Find the words on the list in the word search. Then find four letters in the word search that answer the riddle.

fraction	decimal	terminating	repeating	average	
statistic	convert	numerator	divide	pattern	denominator

D	E	C	I	M	A	L	M	H	T	Y	U	J	K	S	T
I	E	R	E	P	E	A	T	I	N	G	C	F	T	T	E
V	W	N	W	E	R	T	Y	U	I	I	C	D	K	A	R
I	F	V	O	Z	X	C	V	B	N	M	K	U	W	T	M
D	R	Q	E	M	T	G	B	Y	H	N	J	L	Z	I	I
E	A	S	D	U	I	B	G	R	V	F	E	C	I	S	N
P	C	O	L	C	O	N	V	E	R	T	P	O	D	T	A
O	T	M	L	K	I	P	A	V	E	R	A	G	E	I	T
I	I	W	E	R	P	A	T	T	E	R	N	T	R	C	I
U	O	Z	A	Q	W	S	X	C	O	T	G	H	J	I	N
Y	N	G	H	J	K	N	U	M	E	R	A	T	O	R	G

What will the dentist say to you if you do not brush your teeth?

____ ____ ____ ____

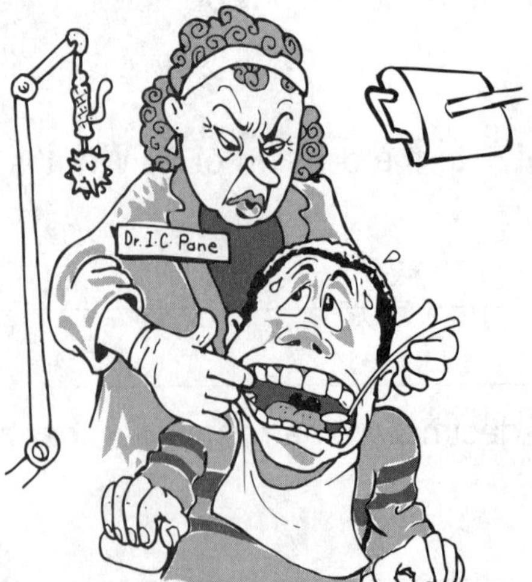

Name ______________________ Date __________ Class __________

LESSON 2-11

Practice A

Comparing and Ordering Rational Numbers

Compare the fractions. Write < or >.

1. $\frac{4}{9} \square \frac{7}{9}$

2. $-\frac{5}{6} \square -\frac{1}{6}$

3. $-\frac{2}{3} \square -\frac{3}{4}$

4. $\frac{12}{5} \square 1\frac{1}{3}$

5. $\frac{4}{7} \square -\frac{6}{7}$

6. $-\frac{1}{4} \square -\frac{1}{7}$

7. $1\frac{5}{8} \square 1\frac{9}{10}$

8. $3\frac{1}{5} \square -3\frac{2}{5}$

Compare the decimals. Write < or >.

9. 7.3 $\square$ 7.6

10. 0.053 $\square$ 0.312

11. −2.51 $\square$ −2.511

12. 1.450 $\square$ 1.405

13. 0.445 $\square$ 0.455

14. $1.\overline{18} \square 1.188$

15. −5.21 $\square$ 5.00

16. $9.\overline{03} \square 9.303$

Write the numbers in order from least to greatest.

17. $0.\overline{5}$, 0.5, $\frac{5}{8}$ ______________

18. 1.34, $1\frac{1}{3}$, 1.3 ______________

19. 2.07, $2\frac{7}{10}$, 2.67 ______________

20. −4.8, −4.08, $-4\frac{1}{8}$ ______________

21. 5.78, $-5\frac{7}{8}$, −5.9 ______________

22. $\frac{6}{7}$, $\frac{5}{6}$, 0.8 ______________

23. $2\frac{2}{3}$, 2.65, 2.69 ______________

24. −0.38, $-\frac{3}{8}$, −0.04 ______________

25. $\frac{3}{7}$, 0.45, $\frac{4}{9}$ ______________

26. Twenty-four karat gold is considered pure. If a ring is 14 karat gold, what is its purity as a fraction and as a decimal rounded to the nearest thousandth?

____

27. Amy scored 9.02 on her balance beam routine. Joella scored 9.2 on her balance beam routine. Which gymnast received the higher score?

____

Name ______________________ Date __________ Class __________

LESSON **2-11**

Practice B

Comparing and Ordering Rational Numbers

Compare the fractions. Write < or >. Justify your answer.

1. $-\frac{7}{8}$ ☐ $-\frac{5}{8}$

2. $\frac{3}{10}$ ☐ $\frac{3}{8}$

3. $4\frac{7}{12}$ ☐ $5\frac{5}{12}$

Compare the decimals. Write < or >. Justify your answer.

4. −0.531 ☐ −0.513

5. 0.73 ☐ 0.073

6. $3.\overline{59}$ ☐ 3.599

Order the numbers from least to greatest.

7. $\frac{4}{9}$, 0.4, 0.45

8. 1.7, 1.65, $1\frac{2}{3}$

9. 3.18, $3\frac{1}{8}$, 3.80

10. −5, −5.25, $-5\frac{2}{5}$

11. $-6\frac{3}{4}$, 6.34, −6.4

12. $\frac{11}{12}$, $\frac{8}{9}$, 0.91

13. $-\frac{3}{5}$, $-\frac{5}{7}$, −0.65

14. 0.3, 0.345, $\frac{1}{3}$

15. −0.75, $\frac{7}{8}$, $-\frac{5}{8}$

16. A ream of paper contains 500 sheets of paper. Norm has 373 sheets of paper left from a ream. Express the portion of a ream Norm has as a fraction and as a decimal.

____

17. The density of Venus, compared to Earth having a density of 1, is 0.943. The density of Mercury is 0.983, compared to the density of Earth. Which planet has a greater density, Venus or Mercury?

____

Name ______________________ Date __________ Class __________

LESSON 2-11

Practice C

Comparing and Ordering Rational Numbers

Compare the fractions. Write < or >. Justify your answer.

1. $-4\frac{3}{4}$ ☐ $-4\frac{5}{6}$

2. $\frac{9}{10}$ ☐ $\frac{10}{11}$

3. $2\frac{2}{5}$ ☐ $2\frac{3}{8}$

Compare the decimals. Write < or >. Justify your answer.

4. 1.031 ☐ 1.3

5. −4.25 ☐ −4.5

6. $2.\overline{17}$ ☐ 2.17

Order the numbers from least to greatest.

7. $\frac{9}{16}$, 0.54, 0.65

8. 3.49, 3.14, $3\frac{4}{9}$

9. −3.05, −3.5, $-3\frac{1}{10}$

10. $-2\frac{7}{12}$, 2.59, −2.7

Choose the greater number.

11. $\frac{4}{5}$ or $\frac{5}{6}$

12. 0.095 or −0.9

13. $-\frac{1}{4}$ or −0.025

14. −5 or −5.01

15. A horsepower is the power needed to lift 33,000 pounds 1 foot in 1 minute. It is equivalent to 746 watts. Express 525 watts as a portion of a horsepower both as a fraction and a decimal rounded to the nearest thousandth.

16. The density of Jupiter, compared to Earth having a density of 1, is 0.24. The density of Saturn is 0.124, compared to the density of Earth. Which planet has a greater density, Jupiter or Saturn?

Name ______________________ Date ____________ Class ____________

LESSON **2-11**

Reteach

Comparing and Ordering Rational Numbers

You can use fraction strips to compare two fractions that have different denominators.

Compare $\frac{2}{3}$ and $\frac{1}{2}$.

$\frac{2}{3}$ is greater than $\frac{1}{2}$.

$\frac{2}{3} > \frac{1}{2}$

$\frac{1}{3}$	$\frac{1}{3}$

$\frac{1}{2}$

Compare $\frac{2}{5}$ and $\frac{3}{4}$.

$\frac{2}{5}$ is less than $\frac{3}{4}$.

$\frac{2}{5} < \frac{3}{4}$

$\frac{1}{5}$	$\frac{1}{5}$

$\frac{1}{4}$	$\frac{1}{4}$	$\frac{1}{4}$

Use fraction strips to compare. Write < or >.

1. $\frac{1}{3}$ ☐ $\frac{1}{2}$

$\frac{1}{3}$

$\frac{1}{2}$

2. $\frac{3}{4}$ ☐ $\frac{5}{8}$

$\frac{1}{4}$	$\frac{1}{4}$	$\frac{1}{4}$

$\frac{1}{8}$	$\frac{1}{8}$	$\frac{1}{8}$	$\frac{1}{8}$	$\frac{1}{8}$

3. $\frac{1}{2}$ ☐ $\frac{3}{5}$ **4.** $\frac{1}{5}$ ☐ $\frac{1}{8}$ **5.** $\frac{3}{8}$ ☐ $\frac{1}{3}$

To compare decimals:

1) Add zeros so that each decimal has the same number of decimal places.
2) Ignore the decimal points and compare the numbers as if they were integers.

Compare 4.3 and 4.27.

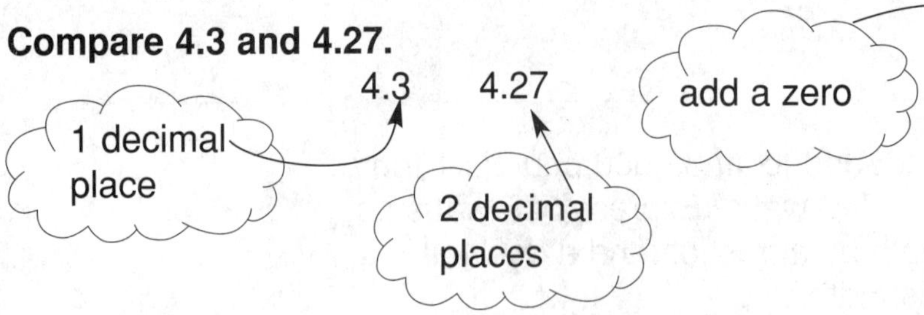

4.30 4.27

430 > 427, so 4.3 > 4.27.

Compare the decimals using < or >.

6. 7.2 ☐ 7.8 **7.** 0.64 ☐ 0.604 **8.** 2.09 ☐ 2.817

Name ______________________ Date ____________ Class ____________

LESSON 2-11

Challenge

Another Rational Number

Between any two rational numbers, there are other rational numbers.

Examples: $\frac{2}{3}$ lies between $\frac{1}{2}$ and $\frac{5}{6}$.

-6 lies between -5.9 and -6.1.

$1\frac{1}{4}$ lies between 1 and $1\frac{1}{2}$.

Write a rational number that lies between each pair of rational numbers below.

1. 0 and $\frac{1}{5}$ ________

2. 3.4 and 3.5 ________

3. 1.08 and 1.8 ________

4. -7 and -7.5 ________

5. -12.1 and -12.2 ________

6. 0 and $-\frac{1}{2}$ ________

7. $1\frac{1}{8}$ and $1\frac{1}{2}$ ________

8. $-2\frac{1}{10}$ and $-2\frac{1}{2}$ ________

9. 3.01 and 3.02 ________

10. 4.2 and $4\frac{1}{2}$ ________

11. -1 and -1.1 ________

12. 3.5 and $3\frac{7}{8}$ ________

13. $-2\frac{3}{8}$ and -2.8 ________

14. 0 and -0.3 ________

15. $7\frac{1}{10}$ and 7.5 ________

16. $5\frac{7}{8}$ and $5\frac{11}{12}$ ________

17. $-\frac{2}{3}$ and $-\frac{3}{4}$ ________

18. $3\frac{1}{5}$ and 3.19 ________

Name ______________________ Date ____________ Class ____________

LESSON **2-11**

Problem Solving

Comparing and Ordering Rational Numbers

Write the correct answer.

1. Some of the wettest summers in the Northeast had the following amounts of precipitation, in inches: $14\frac{7}{10}$, $14\frac{1}{4}$, $14\frac{4}{5}$, and 14.69. Order the amounts of precipitation from least to greatest.

____

____

2. The U.S. soybean crop decreased each year from 2001 to 2003. Production for each year was approximately $2\frac{3}{4}$, 2.9, $2\frac{2}{5}$ billion bushels. Match the soybean crop, in billions of bushels, with its year.

____

____

3. Mark's batting average last year was 0.334. Ken's was 0.304, and Ty's was 0.343. Write the batting averages in order from lowest to highest.

____

4. Kiki lives 0.097 miles from the school, and George lives 0.131 miles from the school. Who lives farthest from the school?

____

Choose the letter for the best answer. Use the graph for problems 5–6.

5. In which year did NFL teams have the greatest success rate?

   A 1992　　**C** 1996

   B 1994　　**D** 2000

6. In which year was the success rate of NFL teams greater than 0.775 and less than 0.78?

   F 1994　　**H** 1997

   G 1995　　**J** 1999

NFL Field Goal Success Rate

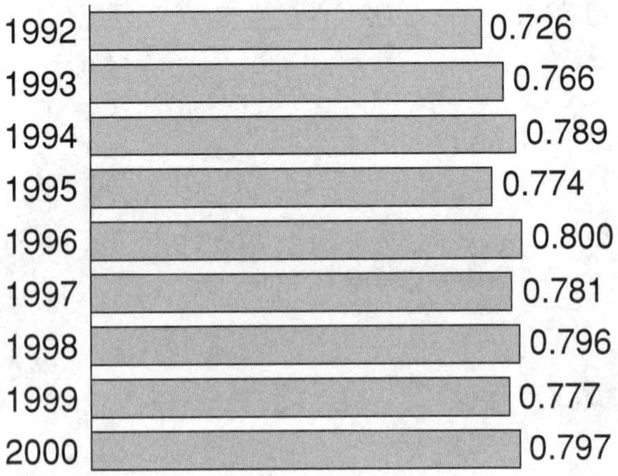

Name ______________________ Date ____________ Class ____________

LESSON **2-11**

Reading Strategies

Compare Visually

You can compare fractions that have the same **denominator** by comparing their **numerators.**

Which is greater, $\frac{4}{9}$ or $\frac{5}{9}$?

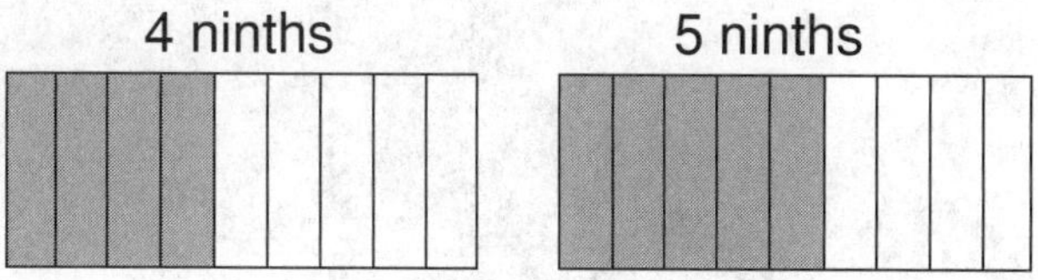

Notice that both fractions are naming ninths.
Look to see which fraction has the most ninths. ⟶ $\frac{5}{9} > \frac{4}{9}$

You can compare fractions visually by using fraction strips.

$\frac{1}{2}$	$\frac{1}{2}$

$\frac{1}{4}$	$\frac{1}{4}$	$\frac{1}{4}$	$\frac{1}{4}$

$\frac{1}{8}$	$\frac{1}{8}$	$\frac{1}{8}$	$\frac{1}{8}$	$\frac{1}{8}$	$\frac{1}{8}$	$\frac{1}{8}$	$\frac{1}{8}$

$\frac{1}{16}$	$\frac{1}{16}$	$\frac{1}{16}$	$\frac{1}{16}$	$\frac{1}{16}$	$\frac{1}{16}$	$\frac{1}{16}$	$\frac{1}{16}$	$\frac{1}{16}$	$\frac{1}{16}$	$\frac{1}{16}$	$\frac{1}{16}$	$\frac{1}{16}$	$\frac{1}{16}$	$\frac{1}{16}$	$\frac{1}{16}$

Compare the following fractions. Use the fraction strips above to help you compare. Write >, <, or =.

1. $\frac{1}{2}$ ☐ $\frac{1}{8}$

2. $\frac{2}{16}$ ☐ $\frac{3}{8}$

3. $\frac{3}{4}$ ☐ $\frac{7}{8}$

4. $\frac{4}{16}$ ☐ $\frac{2}{8}$

5. $\frac{8}{16}$ ☐ $\frac{2}{4}$

6. $\frac{2}{4}$ ☐ $\frac{2}{16}$

Name ______________________ Date __________ Class __________

LESSON **2-11**

Puzzles, Twisters & Teasers

What's the Big Deal?

Use < or > to show which number in each pair is greater. Then use the Answer Key to answer the riddle.

1. 3.8 _____ 3.6
2. 0.088 _____ 0.109
3. −4.26 _____ −4.266
4. −1.902 _____ 0.920
5. 0.755 _____ 0.7
6. 1.82 _____ 1.6
7. 2.25 _____ 2.05
8. 2.88 _____ −2.98
9. 2.05 _____ 2.25
10. 2.88 _____ −2.98
11. 0.32 _____ 0.088
12. 0.999 _____ 1.0
13. −0.93 _____ 0.2
14. 23.44 _____ 23
15. −0.405 _____ −0.45

Answer Key

<	>
O S	I H
T C	A E

What is an insect's least favorite subject in school?

___ ___ ___ ___   ___ ___ ___ ___ ___

< < > >   > < > < <

LESSON 2-1

Practice A
Integers

Graph each integer and its opposite on a number line.

1. 3

−5 −4 −3 −2 −1 0 1 2 3 4 5

2. −5

−5 −4 −3 −2 −1 0 1 2 3 4 5

Use the number line to compare the integers. Write < or >.

−20 −18 −16 −14 −12 −10 −8 −6 −4 −2 0 2 4 6 8 10 12 14 16 18 20

3. −8 [<] 7
4. 4 [>] −7
5. −6 [>] −16
6. −11 [<] 11

Graph the integers on a number line. Then write them in order from least to greatest.

7. −6; 3; −5; 8

−8 −6 −4 −2 0 2 4 6 8

−6; −5; 3; 8

8. 6; −7; −8; 0

−8 −6 −4 −2 0 2 4 6 8

−8; −7; 0; 6

Use a number line to find each absolute value.

−20 −18 −16 −14 −12 −10 −8 −6 −4 −2 0 2 4 6 8 10 12 14 16 18 20

9. |−6| **6**
10. |2| **2**
11. |−1| **1**
12. |8| **8**
13. |−9| **9**
14. |3| **3**
15. |−4| **4**
16. |10| **10**
17. |−15| **15**
18. |20| **20**
19. |−13| **13**
20. |17| **17**

21. The windchill on a cold day made it feel like 5 degrees below zero outside. Write this temperature as an integer.

−5

22. A baby gained 15 pounds from birth to his first birthday. Write this amount as an integer.

15

LESSON 2-1

Practice B
Integers

Graph each integer and its opposite on a number line.

1. 8

−8 −6 −4 −2 0 2 4 6 8

2. −7

−8 −6 −4 −2 0 2 4 6 8

Compare the integers. Use < or >.

3. −15 [<] −7
4. 8 [>] −8
5. −14 [<] 13
6. −18 [>] −20

Use a number line to order the integers from least to greatest.

7. −1; 4; −5; 7; −3

−8 −6 −4 −2 0 2 4 6 8

−5; −3; −1; 4; 7

8. −6; 8; 0; 4; −2

−8 −6 −4 −2 0 2 4 6 8

−6; −2; 0; 4; 8

9. 6; 5; −7; −8; −2

−8 −6 −4 −2 0 2 4 6 8

−8; −7; −2; 5; 6

10. 1; 3; −4; −5; 7

−8 −6 −4 −2 0 2 4 6 8

−5; −4; 1; 3; 7

Use a number line to find each absolute value.

11. |−18| **18**
12. |11| **11**
13. |−25| **25**
14. |19| **19**
15. |−10| **10**
16. |16| **16**
17. |22| **22**
18. |−14| **14**
19. |9| **9**
20. |−24| **24**
21. |−7| **7**
22. |17| **17**

23. Christy dove to a depth of 12 feet below the surface of the water. Write the depth as an integer.

−12

24. The highest point in North Carolina is Mt. Mitchell, with a height of 6,684 feet. Write the height of Mt. Mitchell as an integer.

6,684

LESSON 2-1

Practice C
Integers

Graph each integer and its opposite on a number line.

1. 15

−15 −10 −5 0 5 10 15

2. −4

−5 −4 −3 −2 −1 0 1 2 3 4 5

Compare the integers. Use < or >.

3. −15 [<] 15
4. −7 [>] −9
5. 12 [>] −17
6. −13 [<] 9

Use a number line to order the integers from least to greatest.

7. −2; 3; −6; −1; 4

−8 −6 −4 −2 0 2 4 6 8

−6; −2; −1; 3; 4

8. 7; −5; 6; 0; −7

−8 −6 −4 −2 0 2 4 6 8

−7; −5; 0; 6; 7

Find each absolute value.

9. |−45| **45**
10. |38| **38**
11. |−99| **99**
12. |22| **22**
13. |−19| **19**
14. |375| **375**
15. |−59| **59**
16. |−84| **84**

Compare. Write <, >, or =.

17. |−50| [=] 50
18. 28 [<] |−29|
19. −43 [<] |−37|
20. |48| [<] 53
21. |−79| [>] |65|
22. −92 [<] |−92|
23. |−83| [>] 82
24. |34| [<] |−62|

25. A company reported a net gain of $750,000 for the first quarter and a net loss of $2,000,000 for the second quarter. Write each amount as an integer.

750,000; −2,000,000

26. The record low temperature in Virginia through 1997 was 30°F below 0. The record high temperature was 110°F. Write each temperature as an integer.

−30; 110

LESSON 2-1

Reteach
Integers

This number line shows integers.

0 is neither positive nor negative

negative integers | positive integers

−5 −4 −3 −2 −1 0 1 2 3 4 5

Every integer has an **opposite** integer. A number and its opposite are the same distance from 0.

opposites

4 | 4

−5 −4 −3 −2 −1 0 1 2 3 4 5

1. How many units is 4 from 0? **4**
2. How many units is −4 from 0? **4**
3. 4 and −4 are called **opposite integers**.

Graph each integer and its opposite on a number line.

4. 2

−5 −4 −3 −2 −1 0 1 2 3 4 5

5. −3

−5 −4 −3 −2 −1 0 1 2 3 4 5

You can use a number line to compare and order numbers. The numbers get greater as you move to the right on the number line.

−5 −4 −3 −2 −1 0 1 2 3 4 5

6. What is the order from least to greatest of −1, 2, and −3? **−3; −1; 2**

Write the integers in order from least to greatest.

7. −2; −6; 4

−6; −2; 4

8. −3; 7; 1

−3; 1; 7

The absolute value of an integer is its distance from 0 on a number line. −5 is 5 units from 0. The absolute value of −5 is 5. You write |−5| = 5.

5 units | 5 units

−5 −4 −3 −2 −1 0 1 2 3 4 5

9. How many units from 0 is −3? **3**

LESSON 2-1

Challenge

Order That Integer!

Use two < signs to correctly order each set of integers.

1. −12; 12; −11 — **−12 < −11 < 12**

2. −12; −210; −201 — **−210 < −201 < −12**

3. |−3|; |−6|; |1| — **|1| < |−3| < |−6|**

4. |−38|; |3|; −15 — **−15 < |3| < |−38|**

Use two > signs to correctly order each set of integers.

5. −4; −40; 4 — **4 > −4 > −40**

6. 0; −8; |−18| — **|−18| > 0 > −8**

7. |−75|; |12|; 15 — **|−75| > 15 > |12|**

8. −16; −12; 1 — **1 > −12 > −16**

Find each missing integer. Write its corresponding letter in each blank to answer the riddle.

A	B	C	E	V	T	L	S	U	O	N
−4	−6	7	6	−3	0	−7	−1	1	2	−2

WHAT DID THE CAR DEALER SAY TO THE CUSTOMER?

THIS DEAL IS

A	N
−5; −4; −3	−3; −2; −1

A	B	S	O	L	U	T	E
−4; −3; −2	−7; −6; −5	−1; 0; 1	0; 1; 2	−9; −8; −7	−1; 0; 1	−1; 0; 1	5; 6; 7

V	A	L	U	E
−3; −2; −1	−6; −5; −4	−7; −6; −5	0; 1; 2	6; 7; 8

LESSON 2-1

Problem Solving

Integers

Write the correct answer.

1. The coldest place on record in the United States was in Alaska in 1971. It was 80°F below zero. Write this temperature as an integer.

   −80

2. The temperature outside was −4°F at Jared's house and −8°F at Mario's house. Where was the temperature warmer?

   outside Jared's house

3. A small business reported a net loss of $62,500 during its first year. In its second year, it reported a profit of $34,100. Write each amount as an integer.

   −62,500; 34,100

4. For one day, Lacy recorded the low temperatures in five U.S. cities. The temperatures were 5°C, −1°C, −3°C, 2°C, and 0°C. Write the temperatures in order from least to greatest.

   −3°C; −1°C; 0°C; 2°C; 5°C

Choose the letter for the best answer.

5. Which number is not an integer?

   −3; 5; $\frac{1}{5}$; 0

   A −3 — **(C) $\frac{1}{5}$**

   B 5 — D 0

6. Basha says |5| and |−5| are the same number. Danny says |5| and |−5| are different numbers. Kim says |5| and |−5| both equal 0. Who is correct?

   (F) Basha

   G Danny

   H Kim

   J They are all wrong.

7. Use the table at right. Which continent has the highest point?

   (A) Asia — C Africa

   B South America — D Australia

8. Use the table at right. Which continent has the lowest point?

   F Europe — H North America

   G Australia — **(J) Asia**

Continent	Highest Point (ft)	Lowest Point (ft)
North America	20,320	−282
South America	22,834	−131
Africa	19,340	−512
Asia	29,028	−1,339
Australia	7,310	−52
Europe	18,510	−92

LESSON 2-1

Reading Strategies

Use a Graphic Organizer

Definition	Facts
The set of whole numbers and their opposites	• Each number can be paired with its opposite. The opposite of 2 is −2. The opposite of −3 is 3. • Zero is its own opposite.
Integers	
Examples	**Non-examples**
0, 2, 5, 9, 13, −3, −7, −12, −17	$\frac{2}{3}$, $\frac{11}{5}$, $2\frac{5}{8}$, 0.5, 0.23, 1.05, 3.61

Answer each question.

1. What are integers?

   the set of whole numbers and their opposites

2. Write the opposite of 6. **−6**

3. Write the opposite of 10. **−10**

4. Write the opposite of 0. **0**

5. Write the opposite of −8. **8**

6. Write the opposite of −3. **3**

Write "integer" or "not an integer" for the following numbers.

7. −9 **integer**

8. $\frac{5}{7}$ **not an integer**

9. 0.1 **not an integer**

10. 42 **integer**

LESSON 2-1

Puzzles, Twisters & Teasers

Food for Thought!

Circle words from the list in the word search. Then find a word in the word search that answers the riddle. Write the word on the lines below.

negative, value, opposite, set, number, express, integer, whole, absolute, zero

```
M H T E F V C D E S W I C F R
O P P O S I T E V F R N B H U
X Z O I E S W A X N M T A W H
N E G A T I V E C P U E I H Q
U R A S D F G H V L R G P O K
M O Z X C V B N A R J E L L Y
B P O I U W H O L E K R S E M
E D F G H J K L U W E R T S X
R A B S O L U T E Z A W Q E C
```

What kind of fish goes best with peanut butter?

J E L L Y

LESSON 2-2

Practice A
Adding Integers

Show the addition on the number line. Then write the sum.

1. 2 + (−3)

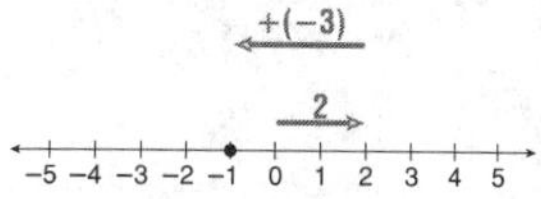

−1

2. −3 + (−4)

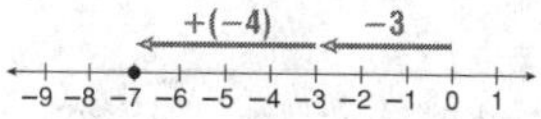

−7

Find each sum.

3. −4 + (−9) **−13**
4. 7 + (−8) **−1**
5. −2 + 1 **−1**
6. 6 + (−9) **−3**
7. 5 + 7 **12**
8. 9 + (−5) **4**
9. (−1) + 9 **8**
10. −9 + (−7) **−16**
11. 2 + (−7) **−5**
12. −6 + (−4) **−10**
13. 3 + 2 **5**
14. −2 + 6 **4**

Evaluate *e* + *f* for the given values.

15. *e* = 9, *f* = −2 **7**
16. *e* = −4, *f* = −6 **−10**
17. *e* = 6, *f* = −1 **5**
18. *e* = −3, *f* = 2 **−1**
19. *e* = 8, *f* = −6 **2**
20. *e* = −2, *f* = −3 **−5**

21. The temperature dropped 13°F in 7 hours. The final temperature was −2°F. What was the starting temperature?

11°F

22. A football team gains 8 yards in one play, then loses 5 yards in the next. What is the team's total yardage?

3 yards

LESSON 2-2

Practice B
Adding Integers

Use a number line to find each sum.

1. −1 + 5

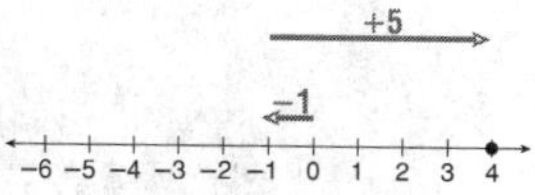

4

2. 4 + (−6)

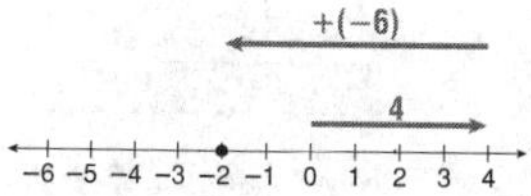

7

Find each sum.

3. −51 + (−9) **−60**
4. 27 + (−6) **21**
5. 1 + (−30) **−29**
6. 15 + (−25) **−10**
7. 50 + (−7) **43**
8. −19 + (−15) **−34**
9. (−23) + 9 **−14**
10. −19 + (−21) **−40**
11. −17 + 11 **−6**
12. 20 + (−8) **12**
13. (−15) + (−7) **−22**
14. 12 + (−14) **−2**

Evaluate *e* + *f* for the given values.

15. *e* = 9, *f* = −24 **−15**
16. *e* = −17, *f* = −7 **−24**
17. *e* = 32, *f* = −19 **13**
18. *e* = −15, *f* = −15 **−30**
19. *e* = −20, *f* = 20 **0**
20. *e* = −30, *f* = 12 **−18**

21. The temperature rose 9°F in 3 hours. If the starting temperature was −5°F, what was the final temperature?

4°F

22. Matt is playing a game. He gains 7 points, loses 10 points, gains 2 points, and then loses 8 points. What is his final score?

−9 points

LESSON 2-2

Practice C
Adding Integers

Use a number line to find each sum.

1. −7 + 10

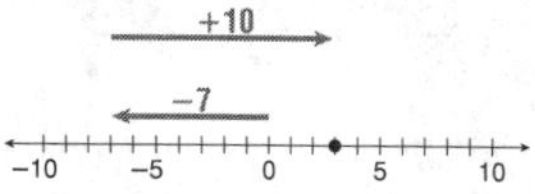

3

2. −2 + (−6)

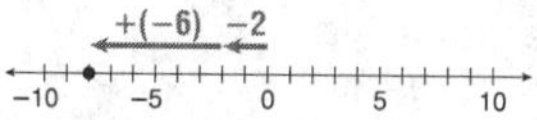

−8

Find each sum.

3. −45 + 9 **−36**
4. 37 + (−60) **−23**
5. −11 + 40 **29**
6. 95 + (−25) **70**
7. 56 + (−37) **19**
8. −79 + (−15) **−94**
9. (−63) + 49 **−14**
10. −49 + (−47) **−96**

Evaluate each expression for *a* = −15, *b* = −8, and *c* = 6.

11. 14 + *b* **6**
12. *c* + (−8) **−2**
13. *a* + *b* **−23**
14. *c* + 5 + *a* **−4**

Evaluate *e* + *f* for the given values.

15. *e* = 90, *f* = −52 **38**
16. *e* = −57, *f* = −23 **−80**
17. *e* = −61, *f* = 32 **−29**

Compare. Write <, >, or =.

18. 25 + (−25) **=** 0
19. 32 + (−19) **>** −51
20. 24 + (−42) **<** 66

21. The temperature dropped 10°F in 5 hours. If the starting temperature was 9°F, what was the final temperature?

−1°F

22. The closing price of a stock gained 2 points on Monday, lost 5 points on Tuesday, lost 1 point on Wednesday, gained 4 points on Thursday, and lost 6 points on Friday. What was the net gain or loss for the week?

−6 points or a loss of 6 points

LESSON 2-2

Reteach
Adding Integers

This balance scale "weighs" positive and negative numbers. Negative numbers go on the left of the balance, and positive numbers go on the right.

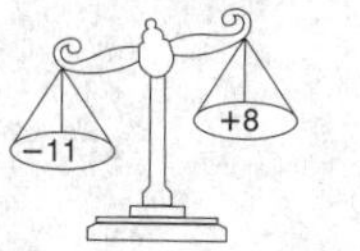

Find −11 + 8.

The scale will tip to the left side because the sum of −11 and +8 is negative.

−11 + 8 = −3

Find −2 + 7.

The scale will tip to the right side because the sum of −2 and +7 is positive.

−2 + 7 = 5

Find −1 + (−3).

Both −1 and −3 go on the left side. The scale will tip to the left side because the sum of −1 and −3 is negative.

−1 + (−3) = −4

Find 3 + (−9).

1. Should you add or subtract? **subtract**
2. Will the sum be positive or negative? **negative**

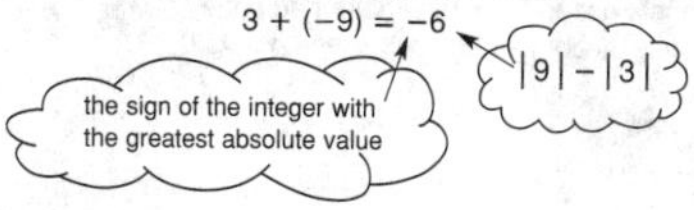

Find −5 + (−8).

3. Should you add or subtract? **add**
4. Will the sum be positive or negative? **negative**
5. −5 + (−8) = **−13**

Add.

6. 7 + −3 = **4**
7. −2 + −3 = **−5**
8. −5 + 4 = **−1**
9. −3 + −1 = **−4**
10. −7 + 9 = **2**
11. 4 + −9 = **−5**

LESSON **2-2**

Challenge

Alphabet Addition

Find the value of each word. Each vowel has a value as shown in the table. All consonants have a value of 2.

A	E	I	O	U
−5	−11	−8	−3	−6

1. ALGEBRA **−13**
2. INTEGER **−22**
3. POSITIVE **−22**
4. NEGATIVE **−27**
5. SIGN **−2**
6. ADDITION **−16**
7. GREATER **−19**
8. LESS **−5**
9. EQUAL **−18**

Write <, >, or = to compare the values of each word pair.

10. MANY **>** FEW
11. ALL **=** NOTHING
12. SCHOOL **>** HOME
13. DOG **>** CAT
14. ADD **<** SUBTRACT
15. STOP **>** GO
16. EVALUATE **<** SOLVE
17. MORE **<** LESS
18. EMPTY **<** FULL
19. Write your name and the first names of 3 other friends or family members. Find the value of each name. Write the names in order from least to greatest value.

    Answers will vary.
20. Write a word with a value of 3.

    Possible answer: chalk
21. Write a word with a value of −3.

    Possible answer: bench
22. What is the highest value word you can think of? Compare it with your classmates' words.

    Answers will vary.

LESSON **2-2**

Problem Solving

Adding Integers

Write the correct answer.

1. The temperature dropped 12°F in 8 hours. If the final temperature was −7°F, what was the starting temperature?

   5°F
2. At 3 P.M., the temperature was 9°F. By 11 P.M., it had dropped 31°F. What was the temperature at 11 P.M.?

   −22°F
3. Tad owes John $23 and borrows $12 more. How much does Tad owe John now?

   $35
4. New Orleans, Louisiana, is 6 feet below sea level. The highest point in Louisiana, Driskill Mountain, is 541 feet higher than New Orleans. How high is Driskill Mountain?

   535 ft
5. A submarine submerged at a depth of −40 ft. dives 57 ft more. What is the new depth of the submarine?

   −97 ft
6. An airplane at 20,000 ft drops 2,500 ft in altitude. What is the new altitude?

   17,500 ft

Choose the letter for the best answer.

7. Last week, Jane made deposits of $64, $25, and $37 into her checking account. She then wrote checks for $52 and $49. What is the overall change in Jane's account balance?

   A −$99   C $126
   (B) $25   D $227
8. In Indianapolis, Indiana, the coldest recorded temperature was −23°F. The hottest recorded temperature was 127°F higher. What was the hottest temperature in Indianapolis?

   F 150°F   **(H) 104°F**
   G 127°F   J −150°F
9. Helena borrowed $189 from her parents to buy an electric bass. She paid back $56 last week and $64 this week. How much does Helena still owe her parents?

   A $133   **(C) $69**
   B $120   D $29
10. The Aral Sea and the Caspian Sea are actually lakes. The elevation of the Caspian Sea is 92 feet below sea level. The Aral Sea is 217 feet higher. What is the elevation of the Aral Sea?

    F −125 ft   H 309 ft
    G −309 ft   **(J) 125 ft**

LESSON **2-2**

Reading Strategies

Use Graphic Aids

Randy's football team had the ball on the zero yard line. On their first play they gained six yards. On the second play they lost four yards. On what yard line is the ball now?

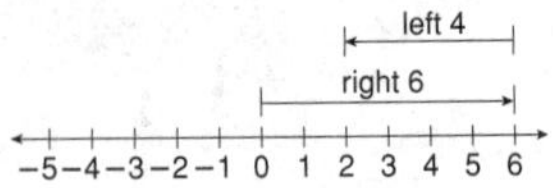

$6 + (-4) = 2$

Use the number line to help you answer the questions.

1. On which number do you begin? **0**
2. Which direction do you move first? How many places do you move?

   to the right; 6
3. Which direction do you move next? How many places do you move?

   to the left; 4

When Angela went to bed, the temperature was zero degrees. When her mother went to bed two hours later, the temperature had gone down 5 degrees. By the time Angela got up the temperature had gone down another 3 degrees. What was the temperature when she got up?

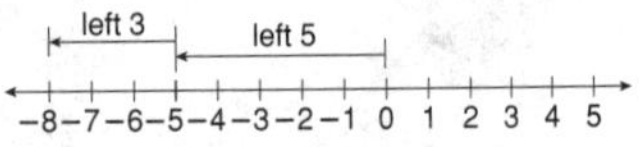

$-5 + (-3) = -8$

Use the number line to help you answer the questions.

4. On which number do you begin? **0**
5. Which direction do you move first? How many spaces?

   to the left; 5
6. Which direction do you move next? How many spaces?

   to the left; 3

LESSON **2-2**

Puzzles, Twisters & Teasers

Chill Out!

Add each pair of integers. Write the letter on the line above the correct answer at the bottom of the page to solve this riddle:

Why did the girl put her money in the freezer?

R	7 + (−16)
S	−3 + 10
L	−6 + (−13)
H	−7 + (−81)
O	−1 + (−5)
E	2 + (−20)
C	25 + 14
W	−3 + 28
D	9 + 4
A	−5 + (−12)
T	6 + (−22)
N	8 + (−7)

S	H	E		W	A	N	T	E	D
7	−88	−18		25	−17	1	−16	−18	13

C	O	L	D		H	A	R	D		C	A	S	H	!
39	−6	−19	13		−88	−17	−9	13		39	−17	7	−88	

LESSON 2-3

Practice A
Subtracting Integers

Show the subtraction on the number line. Then write the difference.

1. 3 − 8

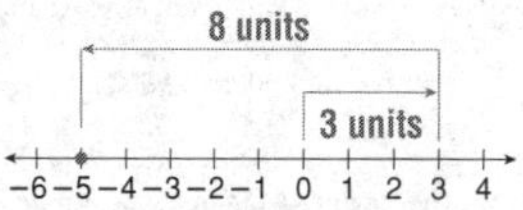

−5

2. −5 − (−1)

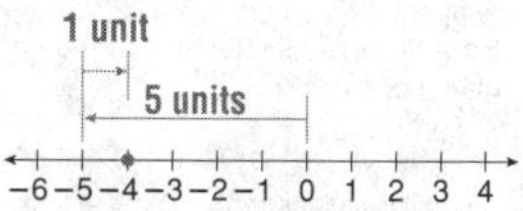

−4

Find each difference.

3. −3 − 4   −7

4. −7 − (−2)   −5

5. 12 − 6   6

6. 2 − (−7)   9

7. −8 − 8   −16

8. −5 − (−5)   0

9. −1 − (−2)   1

10. 9 − (−3)   12

11. 8 − 1   7

12. 7 − (−9)   16

13. −3 − 8   −11

14. −3 − (−7)   4

Evaluate *x* − *y* for each set of values.

15. $x = 6, y = -3$ 9

16. $x = -7, y = 1$ −8

17. $x = -2, y = -5$ 3

18. $x = 9, y = 11$ −2

19. $x = -1, y = -1$ 0

20. $x = -5, y = 5$ −10

21. The high temperature one day was 6°F. The low temperature was −3°F. What was the difference between the high and low temperatures for the day?

9°F

22. The temperature changed from −7°F at 6 A.M. to 7°F at noon. How much did the temperature increase?

14°F

LESSON 2-3

Practice B
Subtracting Integers

Use a number line to find each difference.

1. −2 − 3

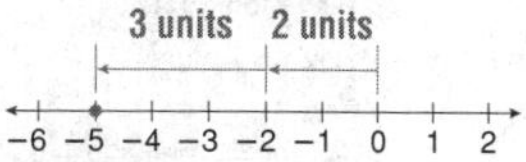

−5

2. 5 − (−1)

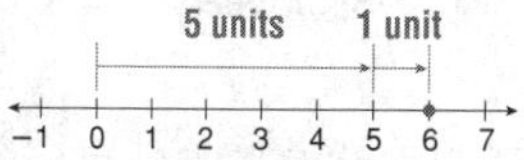

6

Find each difference.

3. −6 − 4   −10

4. −7 − (−12)   5

5. 12 − 16   −4

6. 5 − (−19)   24

7. −18 − (−18)   0

8. 23 − (−23)   46

9. −10 − (−9)   −1

10. 29 − (−13)   42

11. 9 − 15   −6

12. −12 − 14   −26

13. 22 − (−8)   30

14. −16 − (−11)   −5

Evaluate *x* − *y* for each set of values.

15. $x = 14, y = -2$ 16

16. $x = -11, y = 11$ −22

17. $x = -8, y = -15$ 7

18. $x = -9, y = -9$ 0

19. $x = 19, y = -20$ 39

20. $x = 20, y = 25$ −5

21. The high temperature one day was −1°F. The low temperature was −5°F.What was the difference between the high and low temperatures for the day?

4°F

22. The temperature changed from 5°F at 6 P.M. to −2°F at midnight. How much did the temperature decrease?

7°F

LESSON 2-3

Practice C
Subtracting Integers

Use a number line to find each difference.

1. −6 − (−3)

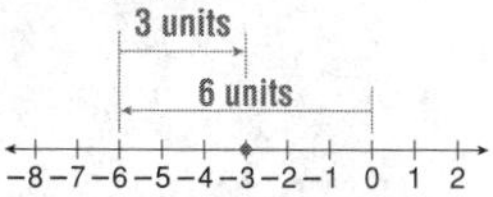

−3

2. −1 − 5

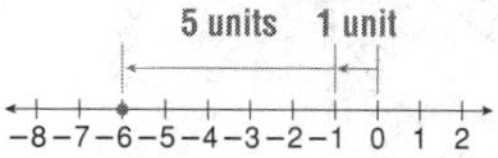

−6

Find each difference.

3. −16 − 15   −31

4. −24 − (−12)   −12

5. 25 − 41   −16

6. 19 − (−19)   38

7. 48 − (−18)   66

8. −37 − 37   −74

9. −50 − (−50)   0

10. −39 − 13   −52

Evaluate *x* − *y* for each set of values.

11. $x = -24, y = 24$ −48

12. $x = -22, y = -32$ 10

13. $x = 18, y = 25$ −7

Simplify.

14. 9 + (−15) − 6   −12

15. 12 − (−5) −3   14

16. −13 − (−8) − (−5)   0

17. −4 −9 + (−7)   −20

18. 10 + (−3) − (−18)   25

19. −25 − 37 − (−5)   −57

20. The record high temperature in North Carolina is 110°F. The record low temperature is −34°F. What is the difference in these temperatures?

144°F

21. The temperature at 6 P.M. was 15°F. By 8 P.M. it had dropped 9 degrees, and then by 10 P.M. it had dropped another 11 degrees. What was the temperature at 10 P.M.?

−5°F

LESSON 2-3

Reteach
Subtracting Integers

The total value of the three cards shown is −6.

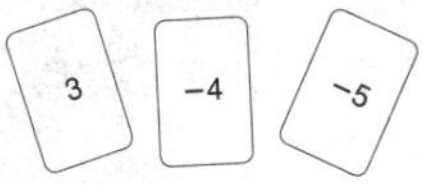

What if you **take away** the 3 card?
Cards −4 and −5 are left.
The new value is −9.
−6 − 3 = −9

What if you **take away** the −4 card?
Cards 3 and −5 are left.
The new value is −2.
−6 − (−4) = −2

Answer each question.

1. Suppose you have the cards shown. The total value of the cards is 12.

a. What if you take away the 7 card? 12 − 7 = 5

b. What if you take away the 13 card? 12 − 13 = −1

c. What if you take away the −8 card? 12 − (−8) = 20

2. Subtract −4 − (−2).

a. −4 < −2. Will the answer be positive or negative? negative

b. |4| − |2| = 2

c. −4 − (−2) = −2

3. Subtract 21 − 13.

a. 21 > 13. Will the answer be positive or negative? positive

b. |21| − |13| = 8

c. 21 − 13 = 8

Subtract.

4. 31 − (− 9) = 40

5. 15 − 18 = −3

6. −9 − 17 = −26

7. −8 − (−8) = 0

8. 29 − (−2) = 31

9. 13 − 18 = −5

LESSON **2-3**

Challenge

Subtract-a-Fact

Complete each temperature and elevation fact. Write the answer in the tables below.

Temperature Facts

1. Louisiana's lowest recorded temperature is (−15 − 1)°F.
2. The highest recorded temperature for Texas is [−18 − (−138)]°F.
3. The normal January temperature for Fairbanks, Alaska, is (−5 − 5)°F.
4. The normal July temperature for Tampa, Florida, is [45 − (−37)]°F.
5. The lowest temperature of 2000 in Detroit, Michigan, was (87 − 90)°F.
6. The highest temperature of 2000 in Norfolk, Virginia, was [−42 − (−138)]°F.
7. North Carolina's lowest recorded temperature is (57 − 91)°F.
8. South Carolina's lowest recorded temperature is (−4 − 15)°F.

Elevation Facts

9. Death Valley's elevation is (−25 − 257) ft.
10. The Caspian Sea's elevation is (125 − 217) ft.
11. Mount Everest's elevation is [−26 − (−29,061)] ft.
12. Mount McKinley's elevation is Mount Everest's elevation − 8,715 ft.
13. Lake Eyre's elevation is (68 − 120) ft.
14. The Dead Sea's elevation is (−761 − 587) ft.
15. Mount Elbrus's elevation is [11,500 − (−7,010)] ft.
16. Lake Assal's elevation is (−254 − 258) ft.

Temperature Facts	°F
Louisiana's lowest	−16°
Texas's highest	120°
Fairbanks normal January	−10°
Tampa normal July	82°
Lowest in Detroit in 2000	−3°
Highest in Norfolk in 2000	96°
North Carolina's lowest	−34°
South Carolina's lowest	−19°

Highest and Lowest Points on Continents	Elevation
Death Valley, North America	−282 ft
Caspian Sea, Europe	−92 ft
Mount Everest, Asia	29,035 ft
Mount McKinley, North America	20,320 ft
Lake Eyre, Australia	−52 ft
Dead Sea, Asia	−1,348 ft
Mount Elbrus, Europe	18,510 ft
Lake Assal, Africa	−512 ft

LESSON **2-3**

Problem Solving

Subtracting Integers

Write the correct answer.

1. The daytime temperature on Mercury can reach 430°C. The nighttime temperature can drop to −180°C. How much can the temperature drop during one day?

   610°C

2. An ice cream company reported a net profit of $24,000 in 2002 and a net loss of $11,000 in 2003. How much did the company's profits change from 2002 to 2003?

   $35,000

3. A small business reported a net loss of $86,000 in 2004 and a net profit of $32,000 in 2005. How much did the company's profits change from 2004 to 2005?

   $118,000

4. The daytime high temperature on the Moon can reach 130°C. At night, the temperature can drop to −110°C. What is the difference between the high and low temperatures?

   240°C

Choose the letter for the best answer.

5. The low point of the Tonga Trench, in the Pacific Ocean, is −10,630 meters. The low point of the Mariana Trench, also in the Pacific Ocean, is 890 meters lower. What is the depth of the Mariana Trench?
   A 10,630 meters
   (B) −11,520 meters
   C −9,740 meters
   D 9,740 meters

6. On Wednesday night in St. Petersburg, Russia, the temperature is −11°C. On the same night in Bombay, India, the temperature is 17°C. What is the difference in temperature?
   F −6°C
   G 50°C
   H −187°C
   (J) 28°C

7. Climax, Colorado, is the highest town in the United States at 11,560 feet. The lowest town is Calipatria, California, which is 185 feet below sea level. What is the difference in elevation?
   A −185 feet
   B 11,375 feet
   C 11,560 feet
   (D) 11,745 feet

8. The low point of the Japanese Trench, in the Pacific Ocean, is −10,372 meters. The low point of the Puerto Rico Trench, in the Atlantic Ocean, is 1,172 meters higher. What is the depth of the Puerto Rico Trench?
   (F) −9,200 meters
   G 8,200 meters
   H −1,172 meters
   J −11,544 meters

LESSON **2-3**

Reading Strategies

Use Graphic Aids

Brett borrowed $7 from his father to buy a CD. He paid back $3. How much money does Brett have now? The number line will help you picture this problem.

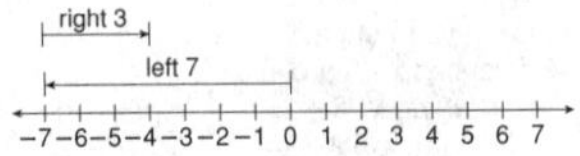

3 − 7 = −4

1. Beginning at 0, in which direction will you move first? **to the left**
2. How many places? **7**
3. Which direction do you move next? **to the right**
4. How many places? **3**
5. On what number do you end? **−4**

Bret does not have any money. He owes his dad $4. He has negative $4.

Sally and her friends made up a game with points. You can either win or lose up to ten points on each round of the game. After the first round, Sally's team had 2 points. In the second round they lost 6 points. How many points was Sally's team down by after the second round?

The number line will help you picture the problem.

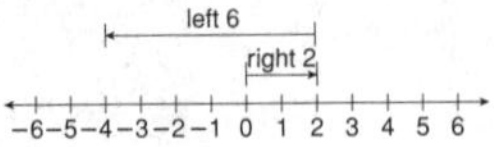

2 − 6 = −4

6. Beginning at zero, which direction will you move first? How many places?

   to the right; 2

7. Which direction will you move next? How many places?

   to the left; 6

8. By how many points was Sally's team down? **4**

LESSON **2-3**

Puzzles, Twisters & Teasers

Integer Invasion!

Subtract each pair of integers. Write the letter on the line above the correct answer at the bottom of the page to answer the following question:

What do you call a pan spinning through space?

C	−12 − (−2)
A	−1 − 8
J	−3 − (−8)
N	6 − 10
B	−31 − 15
U	12 − (−6)
O	−15 − 15
I	30 − (−20)
G	−22 − (−5)
D	−4 − 21
Y	27 − 19
E	−10 − (−7)
R	2 − 9
T	−5 − 9
F	−8 − (−2)

A	N
−9	−4

U	N	I	D	E	N	T	I	F	I	E	D
18	−4	50	−25	−3	−4	−14	50	−6	50	−3	−25

F	R	Y	I	N	G		O	B	J	E	C	T
−6	−7	8	50	−4	−17		−30	−46	5	−3	−10	−14

LESSON 2-4 **Practice A**

Multiplying and Dividing Integers

Find each product.

1. 6 • (−1) −6
2. −4 • 2 −8
3. −3 • (−4) 12
4. −2 • 8 −16
5. 5 • (−7) −35
6. −7 • 9 −63
7. 8 • 4 32
8. −3 • (−5) 15

Multiply.

9. −5 • (−5) 25
10. 8 • (−4) −32
11. −7 • (−6) 42
12. 9 • (−8) −72
13. 1 • (−7) −7
14. −4 • (−5) 20
15. −6 • 3 −18
16. −7 • (−7) 49

Find each quotient.

17. 12 ÷ (−4) −3
18. −15 ÷ (−3) 5
19. −20 ÷ 5 −4
20. −27 ÷ (−9) 3
21. −45 ÷ (−5) 9
22. −18 ÷ 9 −2
23. 24 ÷ (−4) −6
24. 32 ÷ 4 8
25. 21 ÷ 3 7
26. −36 ÷ (−4) 9
27. 16 ÷ (−4) −4
28. −56 ÷ 8 −7
29. −42 ÷ 7 −6
30. −30 ÷ (−6) 5
31. 27 ÷ 9 3
32. 25 ÷ (−5) −5

33. Jenny was walking down a wooded path. For 4 minutes, the elevation dropped steadily. Altogether it dropped 12 feet. What was the change in elevation during the first minute?

−3 feet

34. As a cold front passed, the temperature dropped steadily 2 degrees each hour. It dropped a total of 10 degrees. How many hours did it take for the temperature to drop this far?

5 hours

LESSON 2-4 **Practice B**

Multiplying and Dividing Integers

Find each product.

1. 8 • (−5) −40
2. −4 • 7 −28
3. −6 • (−3) 18
4. −2 • 4 −8
5. 4 • (−9) −36
6. −9 • 5 −45
7. 6 • 8 48
8. −7 • (−3) 21

Multiply.

9. −6 • (−6) 36
10. 9 • (−3) −27
11. −2 • (−8) 16
12. 5 • (−7) −35
13. 10 • 8 80
14. −5 • 9 −45
15. 9 • (−6) −54
16. (−4) • (−11) 44

Find each quotient.

17. 25 ÷ (−5) −5
18. −54 ÷ (−6) 9
19. −10 ÷ 5 −2
20. −28 ÷ (−4) 7
21. −42 ÷ (−7) 6
22. −21 ÷ 3 −7
23. 36 ÷ (−6) −6
24. −81 ÷ (− 9) 9
25. −32 ÷ 8 −4
26. 45 ÷ (−9) −5
27. −72 ÷ (−8) 9
28. 50 ÷ 10 5
29. −42 ÷ 6 −7
30. −72 ÷ (−9) 8
31. 40 ÷ 8 5
32. 56 ÷ (−7) −8

33. Kim was walking down a rocky path. For 4 minutes, the elevation dropped steadily. Altogether it dropped 8 feet. What was the change in elevation per minute for the 4 minutes?

−2 feet

34. As a front passed, the temperature changed steadily over 6 hours. Altogether it changed -18 degrees. What was the change in temperature per hour for the 6 hours?

−3 degrees

LESSON 2-4 **Practice C**

Multiplying and Dividing Integers

Find each product.

1. 9 • (−8) −72
2. −7 • 7 −49
3. −5 • (−6) 30
4. −4 • 8 −32
5. 12 • (−10) −120
6. −50 • 3 −150
7. 8 • 10 80
8. −20 • (−6) 120

Multiply.

9. −7 • (−5) 35
10. 10 • (−9) −90
11. −8 • (−8) 64
12. 12 • (−3) −36
13. −9 • 8 −72
14. 15 • (−4) −60
15. −5 • (−25) 125
16. −150 • 2 −300

Find each quotient.

17. 63 ÷ (−7) −9
18. −48 ÷ (−4) 12
19. −36 ÷ 9 −4
20. −81 ÷ (−9) 9
21. −56 ÷ (−7) 8
22. −54 ÷ 9 −6
23. 80 ÷ (−10) −8
24. −60 ÷ (−12) 5
25. 33 ÷ (−3) −11
26. 100 ÷ (−20) −5
27. −75 ÷ (−3) 25
28. 96 ÷ 8 12
29. −200 ÷ 50 −4
30. −80 ÷ (−4) 20
31. 48 ÷ 12 4
32. 77 ÷ (−11) −7

33. As Randy swam underwater, his elevation changed steadily for 4 minutes. Altogether it changed -20 feet. What was the change in elevation during each minute?

−5 feet

34. Marie withdrew the same amount of money from her savings account each month for 5 months. Altogether she withdrew $300. She did not make any deposits. What was the change in Marie's savings account balance each month?

−$60

LESSON 2-4 **Reteach**

Multiplying and Dividing Integers

Look for the patterns in these products and quotients.

1 • 3 = 3	−1 • 3 = −3	3 ÷ 1 = 3	3 ÷ (−1) = −3
2 • 3 = 6	−2 • 3 = −6	6 ÷ 2 = 3	6 ÷ (−2) = −3
−3 • (−3) = 9	3 • (−3)= −9	−9 ÷ (−3) = 3	−9 ÷ 3 = −3
−4 • (−3) = 12	4 • (−3)= −12	−12 ÷ (−4) = 3	−12 ÷ 4 = −3

Look at how to find the signs of the products.

- The product of two integers with the **same sign** is **positive.**

  (+) • (+) = (+)   (−) • (−) = (+)

- The product of two integers with **different signs** is **negative.**

  (+) • (−) = (−)   (−) • (+) = (−)

Look at how to find the signs of the quotients.

- The quotient of two integers with the **same sign** is **positive.**

  (+) ÷ (+) = (+)   (−) ÷ (−) = (+)

- The quotient of two integers with **different signs** is **negative.**

  (+) ÷ (−) = (−)   (−) ÷ (+) = (−)

Find each product or quotient.

1. −5 • 4 −20
2. 2 • (−8) −16
3. −1 • (−1) 1
4. −6 • 3 −18
5. 7 • (−3) −21
6. −8 • (−4) 32
7. −6 • 5 −30
8. −9 • (−9) 81
9. 36 ÷ (−4) −9
10. −27 ÷ 9 −3
11. −24 ÷ (−6) 4
12. −30 ÷ 5 −6
13. 18 ÷ 6 3
14. 32 ÷ (−8) −4
15. −45 ÷ 9 −5
16. −40 ÷ (−10) 4

LESSON 2-4

Challenge
Football Follies

Phil the Fumbler and Sack'im Sam play for two different football teams. Last year, they both had their worst season ever. Read their interview with a local sports announcer. Find out which player had the worse year by completing the table below.

Announcer: Well Phil, you didn't have a very good season, did you?

Phil: No, I guess losing 42 yards in each of the first 6 games was not what I had hoped to do.

Sam: Gee, that wasn't a very good opening, Phil. At least I gained 53 yards in each of my first 2 games, although, I did lose 88 yards in each of the next 4 games.

Announcer: Yes, Sam, that's true. Now can you tell us how you did in games 7 through 12 last season?

Sam: I can, but I don't really want to.

Announcer: I heard you were sacked and lost 125 yards in each of those games. Is this accurate.

Sam: Yup.

Announcer: Phil, how did you do in games 7 through 12?

Phil: Not that bad. I fumbled quite a bit in games 7 through 11, so I lost 130 yards in each of those games. But I came back in game 12 and gained 85 yards during that one.

Announcer: And what was your yardage for games 13 through 16?

Phil: I lost 95 yards in each game.

Sam: Me, too.

Announcer: Well, I sure hope next season is better for both of you.

Phil and Sam: ME, TOO!

Complete the table.

	Phil	Sam
Total Yards Gained	85 yd	106 yd
Total Yards Lost	−1,282 yd	−1,482 yd
Total Yards	−1,197 yd	−1,376 yd
Average per Game for 16 Games (to the nearest tenth)	−74.8 yd	−86 yd
Which player had the worse year?		✓

LESSON 2-4

Problem Solving
Multiplying and Dividing Integers

Gerald recorded the temperature at 6-hour intervals for one 24-hour period. The chart shows his data.

Time	10:00 P.M.	4:00 A.M.	10:00 A.M.	4:00 P.M.	10:00 P.M.
Temperature (°F)	70	58	64	82	76

Write the correct answer.

1. If the temperature changed steadily, what was the change per hour from 10:00 P.M. the first day to 4:00 A.M.?

   −2°F

2. If the temperature changed steadily, what was the change per hour from 4:00 A.M. to 10:00 A.M.?

   1°F

3. If the temperature changed steadily, what was the change per hour from 10:00 A.M. to 4:00 P.M.?

   3°F

4. If the temperature changed steadily, what was the change per hour from 4:00 P.M. to 10:00 P.M. the second day?

   −1°F

Choose the letter for the best answer.

5. A small company had a profit of −$528 in January. If it continues to have the same profit each month for 4 months, what will be the company's total profit for 4 months?
   - A −$132
   - Ⓒ −$2,112
   - B $132
   - D $2,112

6. Evi's watch is water resistant up to −15 feet. Mateo's watch is water resistant up to 8 times the depth of Evi's watch. Mateo's watch is water resistant to what maximum depth?
   - F −7 feet
   - Ⓗ −120 feet
   - G 120 feet
   - J 7 feet

7. A submarine at −235 meters dives to a depth 8 times its initial depth. To what depth does the submarine dive?
   - A −227 meters
   - B −243 meters
   - Ⓒ −1,880 meters
   - D −8,235 meters

8. A 1-kilogram rock dropped into the ocean would take 37 minutes to reach −6,660 meters. If the rock dropped steadily, how far would it fall in 1 minute?
   - F 167 meters
   - G −167 meters
   - Ⓗ −180 meters
   - J 188 meters

LESSON 2-4

Reading Strategies
Use Graphic Aids

The opposite of 5 is negative 5. Owing money is the opposite of having money. Owing $5 is the opposite of having $5.

1. What is the opposite of owing $10? having $10
2. What is the opposite of having $17? owing $17

David borrowed $4 from his mother each of the last three months. How much money does he owe his mother? The money he owes his mother is a negative number. This problem can be pictured on a number line.

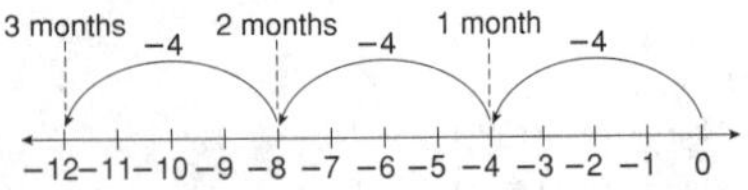

$3 \cdot (-4) = -12$

Use the number line to help you answer the questions.

3. Starting at zero, which direction do you move first? to the left
4. How many places do you move? 4
5. Which direction do you move next? to the left
6. How many places do you move? 4
7. Which direction do you move next? to the left
8. How many places do you move? 4
9. How much money does David owe his mother? $12
10. If David borrowed $4 for one more month, how much would he owe his mother? $16

LESSON 2-4

Puzzles, Twisters & Teasers
It's Raining Money!

Solve the equations. Then use the letter associated with the answer to solve the riddle.

H	$3 \cdot (-3) =$	−9
E	$4 \cdot 2 =$	8
N	$4 \cdot (-2) =$	−8
T	$-4 \cdot 3 =$	−12
I	$3 \cdot 3 =$	9
S	$2 \cdot (-3) =$	−6
C	$-4 \cdot 2 =$	−8
G	$3 \cdot 2 =$	6
A	$-3 \cdot 2 =$	−6
R	$3 \cdot 2 =$	10

When does it rain money?

W H(−9) E(8) N(−8) T(−12) H(−9) E(8) R(10) E(8)

I(9) S(−6) C(−8) H(−9) A(−6) N(−8) G(6) E(8) I(9) N(−8)

T(−12) H(−9) E(8) W E(8) A(−6) T(−12) H(−9) E(8) R(10)

LESSON 2-5
Practice A
Solving Equations Containing Integers

Solve. Check each answer.

1. $n - 6 = -2$ — $n = 4$
2. $x - 8 = -11$ — $x = -3$
3. $7 = a - 5$ — $a = 12$
4. $y + 4 = 2$ — $y = -2$
5. $c + 7 = -3$ — $c = -10$
6. $0 = v + 1$ — $v = -1$
7. $8j = -16$ — $j = -2$
8. $-3k = 24$ — $k = -8$
9. $-20 = -4s$ — $s = 5$
10. $\frac{m}{-2} = -5$ — $m = 10$
11. $\frac{d}{6} = -3$ — $d = -18$
12. $\frac{r}{-7} = 4$ — $r = -28$
13. $p + 8 = -6$ — $p = -14$
14. $-15 = 5b$ — $b = -3$
15. $f - 9 = -1$ — $f = 8$
16. $\frac{n}{6} = -4$ — $n = -24$
17. $k + 10 = 3$ — $k = -7$
18. $4a = -16$ — $a = -4$
19. $-6x = -36$ — $x = 6$
20. $2 = e - 7$ — $e = 9$
21. $3 = \frac{m}{2}$ — $m = 6$

22. The temperature in Minnesota was −8°F one day. This was 12 degrees less than the temperature in Indiana on the same day. What was the temperature in Indiana?

4°F

23. Mr. Harding sold 100 shares of stock at $14 per share. He had a loss of $6 per share. What did Mr. Harding pay for each share of the stock?

$20

LESSON 2-5
Practice B
Solving Equations Containing Integers

Solve. Check each answer.

1. $y - 5 = -4$ — $y = 1$
2. $n - 9 = -14$ — $n = -5$
3. $13 = x - 15$ — $x = 28$
4. $p + 18 = 14$ — $p = -4$
5. $q + 6 = -2$ — $q = -8$
6. $0 = w + 4$ — $w = -4$
7. $9h = -36$ — $h = -4$
8. $-3b = 36$ — $b = -12$
9. $-100 = -4u$ — $u = 25$
10. $\frac{d}{5} = -7$ — $d = -35$
11. $\frac{c}{4} = -20$ — $c = -80$
12. $\frac{s}{-9} = 9$ — $s = -81$
13. $f + 15 = -16$ — $f = -31$
14. $-75 = 3v$ — $v = -25$
15. $g - 19 = -21$ — $g = -2$
16. $-63 = -9s$ — $s = 7$
17. $14 + m = -10$ — $m = -24$
18. $12 = \frac{w}{4}$ — $w = 48$
19. $x = 15 - 31$ — $x = -16$
20. $\frac{e}{-7} = 8$ — $e = -56$
21. $-6 = 21 - n$ — $n = 27$

22. The temperature in Buffalo, New York, was −2°F one day. This was 42 degrees warmer than the temperature in Nome, Alaska, on the same day. What was the temperature in Nome?

−44°F

23. LaSanda bought 20 shares of stock for $175. She sold the stock for a total profit of $25. What was the selling price of each share of stock?

$10

LESSON 2-5
Practice C
Solving Equations Containing Integers

Solve. Check each answer.

1. $h - 15 = -17$ — $h = -2$
2. $a - 29 = -12$ — $a = 17$
3. $13 = g - 20$ — $g = 33$
4. $m + 48 = 31$ — $m = -17$
5. $n + 26 = -4$ — $n = -30$
6. $0 = z + 50$ — $z = -50$
7. $15y = -75$ — $y = -5$
8. $0 = -7d$ — $d = 0$
9. $-144 = -8v$ — $v = 18$
10. $\frac{k}{-5} = -14$ — $k = 70$
11. $\frac{r}{4} = -250$ — $r = -1{,}000$
12. $\frac{t}{-11} = 11$ — $t = -121$
13. $s + 25 = -18$ — $s = -43$
14. $-96 = -4j$ — $j = 24$
15. $x - 49 = -57$ — $x = -8$

Translate each sentence into an equation. Then solve the equation.

16. 8 less than a number n is −13. — $n - 8 = -13$; $n = -5$
17. A number f multiplied by −12 is −72. — $-12f = -72$; $f = 6$
18. The sum of −47 and a number d is 84. — $-47 + d = 84$; $d = 131$
19. 8 less than a number n is −13. — $8q = -40$; $q = -5$

20. Jake bought 50 shares of stock for $1,000 and sold them for a loss of $375. What was the selling price of each share of stock?

$12.50

21. Mrs. Lee bought 50 shares of Wiget stock. She paid a total of $1,200, which included a $25 commission fee. Two years later she sold all 50 shares for a total of $1,500. This included a $25 commission fee to the broker. What was the cost and selling price per share of the stock?

cost: $23.50; selling price: $29.50

LESSON 2-5
Reteach
Solving Equations Containing Integers

- You can use addition to solve an equation involving subtraction. Addition undoes subtraction. Adding the same number to both sides of the equation keeps the equation balanced.

	Check
$x - 5 = -6$	$x - 5 = -6$
$x - 5 + 5 = -6 + 5$	$-1 - 5 \stackrel{?}{=} -6$
$x = -1$	$-6 \stackrel{?}{=} -6$ ✓

- You can use subtraction to solve an equation involving addition. Subtraction undoes addition. Subtracting the same number from both sides of the equation keeps the equation balanced.

	Check
$n + 4 = -15$	$n + 4 = -15$
$n + 4 - 4 = -15 - 4$	$-19 + 4 \stackrel{?}{=} -15$
$n = -19$	$-15 \stackrel{?}{=} -15$ ✓

Solve. Check your answer.

1. $p - 9 = -3$
   $p - 9 + \underline{9} = -3 + \underline{9}$
   $p = 6$
2. $w - 2 = -14$
   $w - 2 + \underline{2} = -14 + \underline{2}$
   $w = -12$
3. $x - 12 = -5$
   $x - 12 + \underline{12} = -5 + \underline{12}$
   $x = 7$
4. $f - 8 = 6$
   $f - 8 + \underline{8} = 6 + \underline{8}$
   $f = 14$
5. $6 = m - 7$ — $m = 13$
6. $-4 = s - 10$ — $s = 6$
7. $-8 = y - 2$ — $y = -6$
8. $a + 19 = 7$ — $a = -12$
9. $b + 15 = -9$ — $b = -24$
10. $39 + t = 45$ — $t = 6$
11. $-5 = x + 7$ — $x = -12$
12. $-2 = k + 11$ — $k = -13$
13. $10 = -3 + j$ — $j = 13$

LESSON **2-5**

Reteach

Solving Equations Containing Integers (continued)

- You can use division to solve an equation involving multiplication. Division undoes multiplication. Dividing both sides of the equation by the same number keeps the equation balanced.

	Check
$3y = -9$	$3y = -9$
$\frac{3y}{3} = \frac{-9}{3}$	$3 \cdot (-3) \stackrel{?}{=} -9$
$y = -3$	$-9 \stackrel{?}{=} -9$ ✓

- You can use multiplication to solve an equation involving division. Multiplication undoes division. Multiplying both sides of an equation by the same number keeps the equation balanced.

	Check
$\frac{a}{-5} = -8$	$\frac{a}{-5} = -8$
$-5 \cdot \frac{a}{-5} = -8 \cdot (-5)$	$\frac{40}{-5} \stackrel{?}{=} -8$
$a = 40$	$-8 \stackrel{?}{=} -8$ ✓

Solve. Check your answer.

14. $5g = -35$

$\frac{5g}{5} = \frac{-35}{5}$

$g = -7$

15. $-8y = -96$

$\frac{-8y}{-8} = \frac{-96}{-8}$

$y = 12$

16. $54 = -6f$

$\frac{54}{-6} = \frac{-6f}{-6}$

$f = -9$

17. $3e = -33$

$e = -11$

18. $-49 = 7n$

$n = -7$

19. $-75 = -5c$

$c = 15$

20. $\frac{n}{4} = -15$

$n = -60$

21. $\frac{m}{-6} = -9$

$m = 54$

22. $\frac{s}{-10} = 8$

$s = -80$

23. $4 = \frac{w}{-6}$

$w = -24$

24. $9 = \frac{z}{5}$

$z = 45$

25. $-11 = \frac{h}{6}$

$h = -66$

LESSON **2-5**

Challenge

Balloon Trail

Match each statement with one of the equations A through J. Write the letter of the equation in the space provided. Then solve each equation and write the solution. To find the way to the party, draw a path through the balloons that show the correct solutions to Exercises 1 to 10 in order.

A. $-7 + x = -3$ **B.** $\frac{n}{4} = -8$ **C.** $3y = -36$ **D.** $x - 9 = -4$

E. $-6n = 42$ **F.** $\frac{y}{5} = 10$ **G.** $x + 12 = -5$ **H.** $n - 6 = 2$

I. $-9y = -27$ **J.** $\frac{y}{-8} = -2$

1. Three times a number is −36. C; $y = -12$
2. The sum of a number and 12 is −5. G; $x = -17$
3. A number decreased by 9 is −4. D; $x = 5$
4. A number divided by 4 is −8. B; $n = -32$
5. A number divided by 5 is 10. F; $y = 50$
6. A number divided by −8 is −2. J; $y = 16$
7. 6 less than a number is 2. H; $n = 8$
8. The product of a number and −6 is 42. E; $n = -7$
9. Add −7 to a number. The sum is −3. A; $x = 4$
10. −9 times a number is −27. I; $y = 3$

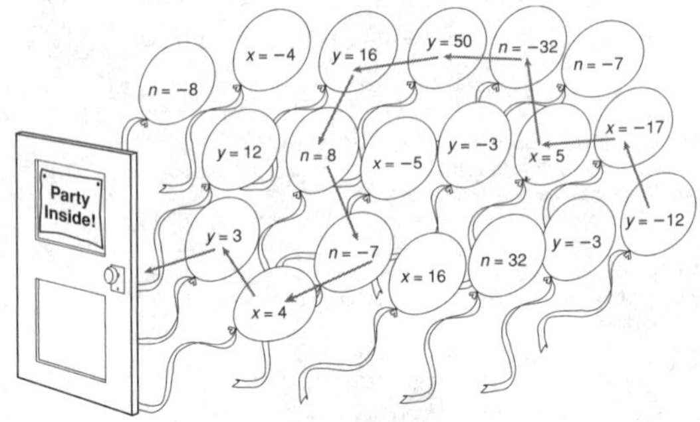

LESSON **2-5**

Problem Solving

Solving Equations Containing Integers

Write the correct answer.

1. Jolene has 30 days to prepare for a bicycle race. She will bicycle 15 miles each day. How many miles will Jolene have bicycled during her training?

   450 mi

2. When the amount of money spent is greater than the amount of income, it is called a deficit. In 2002, the U.S. budget deficit was −$158 billion. By 2004, it was −$412 billion. How much did the deficit go up during those two years?

   $254 billion

3. Two angles of a triangle have a sum of 110°. The sum of all three angles is 180°. What is the measure of the third angle?

   70°

4. The Statue of Liberty was erected in New York City in 1886. How old is the statue?

   Possible answer: 118 years old (in 2004)

Choose the letter for the best answer.

5. Lonnie swam 3 miles each day for 45 days. How many miles did Lonnie swim?

   (A) 135 miles
   B 48 miles
   C 125 miles
   D 45 miles

6. On the hottest day in Richmond, Virginia, the temperature was 105°F. On the coldest day, the temperature was 117°F lower. What was the coldest temperature in Richmond?

   F −117°F
   (G) −12°F
   H 12°F
   J 117°F

7. Mr. Marco sold 50 shares of Gizmo stock for $1,250. What was the selling price of the stock per share?

   A $1,200
   B $250
   C $50
   (D) $25

8. The U.S. Bullion Depository at Fort Knox contains about 147 million troy ounces of gold. At a market price of $400 per troy ounce, what is the approximate value of the gold stored at Fort Knox?

   F $6 billion
   (G) $60 billion
   H $560 million
   J $560 billion

LESSON **2-5**

Reading Strategies

Use a Flowchart

The rules for solving equations with integers are the same as with whole numbers.

Use a flowchart to help you follow the rules.

1		2		3		4
Get the variable by itself.	→	Perform the same operation on both sides of the equation.	→	Apply rules for computing integers.	→	Check the solution.

$x + 12 = 5$ → Get x by itself.

$x + 12 - \mathbf{12} = 5 - \mathbf{12}$ → Subtract 12 from both sides.

$x = \mathbf{5 + (-12)}$ → Add the opposite.

$x = -7$ → Check.

$x + 12 = 5$

$-7 + 12 \stackrel{?}{=} 5$

$5 = 5$

Use $w - 12 = (-4)$ to answer Exercises 1–4.

1. What operation is used in this equation?

   subtraction

2. What operation will you use to get the variable by itself?

   addition

3. Apply this operation to both sides of the equation.

   $w - 12 + 12 = -4 + 12$

4. What is the value of w?

   $w = 8$

Use $x + (-9) = (-4)$ to answer Exercises 5 and 6.

5. What operation is used in this problem?

   addition

6. What operation will you use to get x by itself?

   subtraction

LESSON **2-5**

Puzzles, Twisters & Teasers

A Bold Conclusion!

Solve each equation. Then use the boldfaced letters to answer the riddle.

Q	$9y = 900$	$y = 100$
A	$d - 15 = 45$	$d = 60$
U	$j + 56 = -7$	$j = 63$
C	$v - 39 = -16$	$v = 23$
F	$w + 41 = 0$	$w = -41$
K	$-85 = -5c$	$c = 17$
X	$11y = -121$	$y = -11$
E	$-23 = x + 35$	$x = -58$
O	$200 = -25p$	$p = -8$
R	$-20 = 2z$	$z = -10$
B	$a - 25 = -5$	$a = 20$
S	$s + 2 = -4$	$s = -6$
W	$-5a = 300$	$a = -60$
D	$k = -18 + 11$	$k = -7$
G	$b - 7 = -16$	$b = -9$

What do you call ten ducks in a cardboard box?

A (60) B (20) O (−8) X (−11) O (−8) F (−41)

Q (100) U (−63) A (60) C (23) K (17) E (−58) R (−10) S (−6)

LESSON **2-6**

Practice A

Prime Factorization

Tell whether each number is prime or composite.

1. 25 — Prime
2. 16 — Composite
3. 21 — Composite
4. 29 — Prime

Choose the letter for the best answer.

5. What is the prime factorization of 24?
   - (A) $2^3 \cdot 3$
   - B 3
   - C $2^2 \cdot 3^2$
   - D $2^3 \cdot 3^2$

6. What is the prime factorization of 50?
   - F $3 \cdot 5^2$
   - G $3^2 \cdot 5$
   - (H) $2 \cdot 5^2$
   - J

7. What is the prime factorization of 35?
   - A $2 \cdot 3 \cdot 5$
   - (B) $5 \cdot 7$
   - C $3^2 \cdot 5$
   - D $3^2 \cdot 7$

8. What is the prime factorization of 27?
   - (F) 3^3
   - G
   - H 9^2
   - J $2^2 \cdot 3^2$

Use a factor tree to find the prime factorization.

9. 25 — 5^2
10. 20 — $2^2 \cdot 5$
11. 32 — 2^5
12. 49 — 7^2
13. 30 — $2 \cdot 3 \cdot 5$
14. 44 — $2^2 \cdot 11$
15. 63 — $3^2 \cdot 7$
16. 28 — $2^2 \cdot 7$

Use a step diagram to find the prime factorization.

17. 93 — $3 \cdot 31$
18. 110 — $2 \cdot 5 \cdot 11$
19. 80 — $2^4 \cdot 5$
20. 42 — $2 \cdot 3 \cdot 7$
21. 625 — 5^4
22. 98 — $2 \cdot 7^2$
23. 140 — $2^2 \cdot 5 \cdot 7$
24. 92 — $2^2 \cdot 23$

LESSON **2-6**

Practice B

Prime Factorization

Tell whether each number is prime or composite.

1. 33 — Composite
2. 41 — Prime
3. 52 — Composite
4. 79 — Prime
5. 96 — Composite
6. 121 — Composite
7. 83 — Prime
8. 119 — Composite

Write the prime factorization of each number.

9. 57 — $3 \cdot 19$
10. 49 — 7^2
11. 88 — $2^3 \cdot 11$
12. 95 — $5 \cdot 19$
13. 105 — $3 \cdot 5 \cdot 7$
14. 98 — $2 \cdot 7^2$
15. 52 — $2^2 \cdot 13$
16. 42 — $2 \cdot 3 \cdot 7$
17. 68 — $2^2 \cdot 17$
18. 91 — $7 \cdot 13$
18. 60 — $2^2 \cdot 3 \cdot 5$
20. 72 — $2^3 \cdot 3^2$
21. 56 — $2^3 \cdot 7$
22. 144 — $2^4 \cdot 3^2$
23. 370 — $2 \cdot 5 \cdot 37$
24. 168 — $2^3 \cdot 3 \cdot 7$
25. 124 — $2^2 \cdot 31$
26. 515 — $5 \cdot 103$
27. 725 — $5^2 \cdot 29$
28. 220 — $2^2 \cdot 5 \cdot 11$
29. 450 — $2 \cdot 3^2 \cdot 5^2$
30. 1,000 — $2^3 \cdot 5^3$
31. 1,040 — $2^4 \cdot 5 \cdot 13$
32. 2,500 — $2^2 \cdot 5^4$

33. The prime factorization of a number is $3^2 \cdot 5 \cdot 11$. What is the number?

    495

LESSON **2-6**

Practice C

Prime Factorization

Tell whether each number is prime or composite.

1. 91 — Prime
2. 117 — Composite
3. 109 — Prime
4. 73 — Prime

Write the prime factorization of each number.

5. 84 — $2^2 \cdot 3 \cdot 7$
6. 343 — 7^3
7. 135 — $3^3 \cdot 5$
8. 180 — $2^2 \cdot 3^2 \cdot 5$

Use a step diagram to find the prime factorization.

9. 290 — $2 \cdot 5 \cdot 29$
10. 900 — $2^2 \cdot 3^2 \cdot 5^2$
11. 575 — $5^2 \cdot 23$
12. 90 — $2 \cdot 3^2 \cdot 5$
13. 220 — $2^2 \cdot 5 \cdot 11$
14. 350 — $2 \cdot 5^2 \cdot 7$
15. 495 — $3^2 \cdot 5 \cdot 11$
16. 480 — $2^5 \cdot 3 \cdot 5$
17. 2,000 — $2^4 \cdot 5^3$
18. 1,040 — $2^4 \cdot 5 \cdot 13$
19. 1,650 — $2 \cdot 3 \cdot 5^2 \cdot 11$
20. 3,000 — $2^3 \cdot 3 \cdot 5^3$

21. The prime factorization of a number is $2^2 \cdot 3^5$. What is the number?

    972

22. You can factor 150 as $15 \cdot 10$. List four other ways can you factor 150.

    $5 \cdot 30$; $3 \cdot 50$; $6 \cdot 25$; $2 \cdot 75$

23. A number x is a prime factor of 65 and 104. What is the number?

    13

Write the composite number for each prime factorization.

24. $2^5 \cdot 5^2$ — 800
25. $3^3 \cdot 5 \cdot 11$ — 1,485
26. $4^3 \cdot 7 \cdot 13$ — 5,824
27. $2^8 \cdot 3^5$ — 62,208

28. The prime factors of a number are all the prime numbers between 20 and 30. No factor is repeated. What is the number?

    667

LESSON **2-6**

Reteach
Prime Factorization

To write the **prime factorization** of a number, write the number as the product of only prime numbers. A prime number has only two factors, itself and 1.

Use a *factor tree* to find the prime factorization of 18.

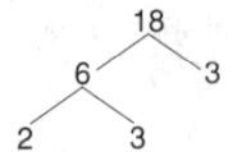

Keep factoring until all the factors are prime.

Use a *factor tree* to find the prime factorization of 36.

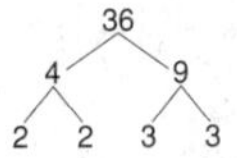

The prime factorization of 18 is $2 \cdot 3 \cdot 3$, or $2 \cdot 3^2$.

The prime factorization of 36 is $2 \cdot 2 \cdot 3 \cdot 3$, or $2^2 \cdot 3^2$.

Use a factor tree to find the prime factorization.

1. 20 $2^2 \bullet 5$
2. 28 $2^2 \bullet 7$
3. 54 $2 \bullet 3^3$
4. 63 $3^2 \bullet 7$

Use a *step diagram* to find the prime factorization of 60.

2|60
2|30
3|15
5|5
1

1) The divisors must be prime numbers.
2) Keep dividing until the quotient is 1.
3) The divisors are the factors in the prime factorization.

Use a *step diagram* to find the prime factorization of 75.

3|75
5|25
5|5
1

The prime factorization of 60 is $2 \cdot 2 \cdot 3 \cdot 5$, or $2^2 \cdot 3 \cdot 5$.

The prime factorization of 75 is $3 \cdot 5 \cdot 5$, or $3 \cdot 5^2$.

Use a step diagram to find the prime factorization.

5. 48 $2^4 \bullet 3$
6. 24 $2^3 \bullet 3$
7. 40 $2^3 \bullet 5$
8. 98 $2 \bullet 7^2$

LESSON **2-6**

Challenge
A Different Point of View

Create two more factor trees with different factors for each number. **Possible answers are given.**

1.

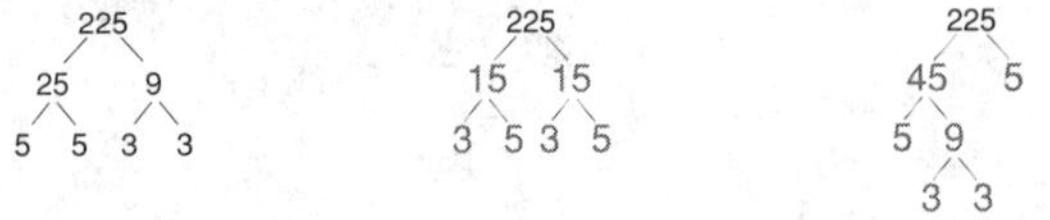

Prime factorization of 225: $3^2 \bullet 5^2$

2.

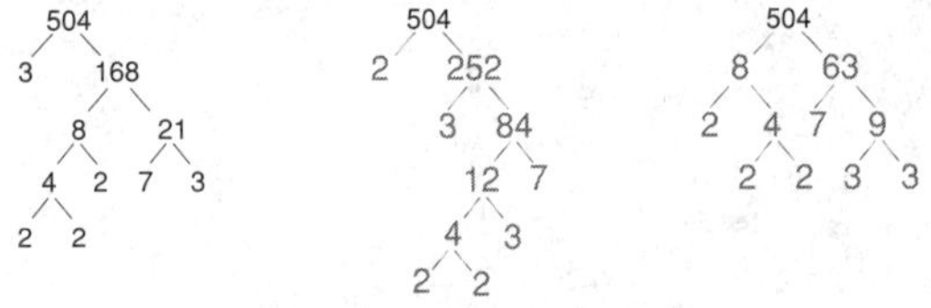

Prime factorization of 504: $2^3 \bullet 3^2 \bullet 7$

3. In 1742, Christian Goldbach wrote his now famous Goldbach's Conjecture, which states that every even number greater than 2 can be represented as the sum of 2 primes. Although this conjecture is still an open question, it can be supported. Choose 3 even numbers greater than 2 to help support Goldbach's Conjecture.

   Possible answer: $12 = 7 + 5$; $32 = 19 + 13$; $44 = 13 + 31$

4. Goldbach also made a conjecture that every odd number is the sum of 3 primes. Choose 3 odd numbers to support this conjecture.

   Possible answer: $13 = 3 + 5 + 5$; $15 = 5 + 5 + 5$; $33 = 23 + 5 + 5$

LESSON **2-6**

Problem Solving
Prime Factorization

Write the correct answer.

1. The width of a swimming pool (in feet) is a prime number greater than 10. The width and length of the pool are factors of 408. What are the dimensions of the pool?

   17 ft by 24 ft

2. The area of the dining room at Thomas Jefferson's home in Monticello is about 342 square feet. If the approximate length of one side is a prime number less than 25, what are the approximate dimensions of the room?

   19 ft by 18 ft

3. A university has a lounge that can be converted into a meeting hall for 250 people. If the hall is filled and everyone is in equal groups, what are the different ways the people can be grouped so that there are no more than 10 groups?

   5 groups of 50; 2 groups of 125; 10 groups of 25

4. You want to read a mystery that has 435 pages. If you read the same number of pages per day and the number is a prime number greater than 20, how many pages per day will you read?

   29 pages per day for 15 days

Choose the letter of the best answer.

5. There are 228 seventh graders. Each seventh-grade homeroom starts the year with the same number of students and has at least 15 students. What is the least number of rooms that are needed?

   A 17
   B 19
   (C) 12
   D 21

6. Solve this riddle: I am a number whose prime factors are all the prime numbers between 6 and 15. No factor is repeated. What number am I?

   F 9,009
   G 91
   (H) 1,001
   J 6,006

7. What is the prime factorization of 1,485?

   (A) $3 \cdot 3 \cdot 3 \cdot 5 \cdot 11$
   B $3 \cdot 3 \cdot 5 \cdot 5 \cdot 11$
   C $3 \cdot 5 \cdot 9 \cdot 11$
   D $5 \cdot 11 \cdot 27$

8. Solve this riddle: I am a prime factor of 39 and 65. What number am I?

   F 3
   G 5
   H 11
   (J) 13

LESSON **2-6**

Reading Strategies
Compare and Contrast

Whole numbers are either **prime numbers** or **composite numbers.** Compare the differences between them.

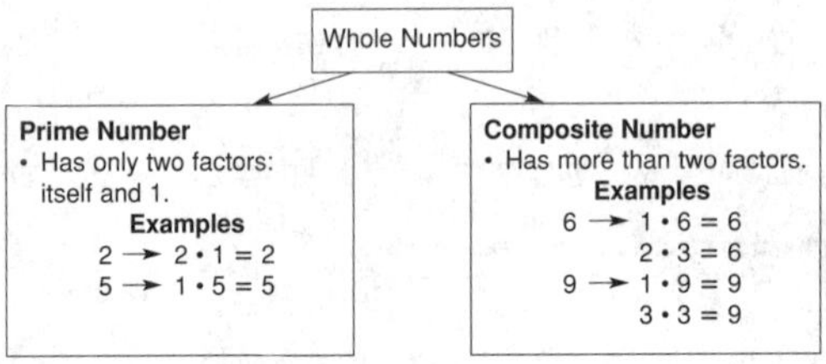

A **composite number** can be written as the product of prime numbers. This is called **prime factorization.**
A **factor tree** helps you find all prime factors of a number.

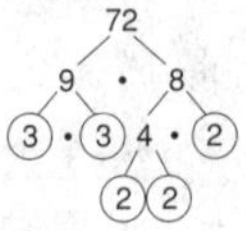

3 is a prime number. Circle both 3's.
2 is a prime number. Circle both 2's.
9 and 8 are not prime numbers. Continue to factor.
2 is a prime number. Circle 2.
4 is not a prime number. Continue to factor.

The prime factorization of 72 is $3 \cdot 3 \cdot 2 \cdot 2 \cdot 2$, or $3^2 \cdot 2^3$.

Answer each question.

1. What is the difference between a prime number and a composite number?

   A prime number has only two factors, itself and 1. A composite number has more than two factors.

2. Find the prime factorization of 48. **$2 \bullet 2 \bullet 2 \bullet 2 \bullet 3$, or $2^4 \bullet 3$**
3. Find the prime factorization of 56. **$2 \bullet 2 \bullet 2 \bullet 7$, or $2^3 \bullet 7$**

LESSON **2-6**

Puzzles, Twisters & Teasers

Can You Find All the Words?

Solve the word search, then find the word from the list that best completes the riddle.

step	unique	composite	factor	factorization
exponent	tree	prime	diagram	Hajratwala

P R I M E F U N I Q U E P
C W E S A H D C B N U O F
T Y P L M A S X D S R W K
E R T Y U J C F T S Y C B
F A C T O R I Z A T I O N
A S Z R A A Q W R E I M F
C G B E X T D S A P B P M
T N M E A W N J I L K O C
O W E R T A T H N B V S I
R Y U I O L C V B N M I L
A S E D I A G R A M F T W
W S C E X P O N E N T E J

What was the mathematician's favorite food?

P R I M E rib

LESSON **2-7**

Practice A

Greatest Common Factor

Draw a line to the greatest common factor (GCF). For Exercises 1–10, choose from A–K; for Exercises 11–20, choose from M–W.

1. 21, 35	**A.** 3	**11.** 18, 48	**M.** 16
2. 12, 32	**B.** 7	**12.** 10, 25	**N.** 60
3. 27, 66	**C.** 5	**13.** 120, 180	**P.** 6
4. 6, 12, 30	**D.** 4	**14.** 16, 48, 80	**Q.** 12
5. 20, 45, 75	**E.** 9	**15.** 6, 16, 30	**R.** 5
6. 27, 36, 45	**F.** 6	**16.** 60, 84, 120	**S.** 3
7. 25, 75, 100	**G.** 19	**17.** 42, 63, 84	**T.** 9
8. 38, 57	**H.** 8	**18.** 64, 96	**U.** 2
9. 32, 40, 120	**J.** 25	**19.** 18, 99, 135	**V.** 21
10. 36, 48, 60	**K.** 12	**20.** 24, 45, 90	**W.** 32

21. A jewelry designer is making identical bracelets. She has 21 blue beads and 14 red beads. What is the greatest number of bracelets she can make using all of the beads?

7 bracelets

22. A florist is preparing identical flower baskets for a banquet. He has 72 tulips and 60 roses. What is the greatest number of flower baskets he can prepare using all of the tulips and roses?

12 flower baskets

LESSON **2-7**

Practice B

Greatest Common Factor

Find the greatest common factor (GCF).

1. 12, 15 — 3
2. 22, 33 — 11
3. 63, 45 — 9
4. 15, 50 — 5
5. 18, 81 — 9
6. 18, 48 — 6
7. 20, 24 — 4
8. 14, 42, 49 — 7
9. 3, 6, 9 — 3
10. 16, 24, 30 — 2
11. 16, 40, 88 — 8
12. 42, 70 — 14
13. 25, 125, 200 — 25
14. 26, 39, 52 — 13
15. 36, 100 — 4
16. 35, 77 — 7
17. 56, 84 — 28
18. 14, 49, 56, 84 — 7
19. 30, 75, 60, 90 — 15
20. 12, 38, 40, 94 — 2
21. 48, 66, 96, 102 — 6

22. Volunteers are preparing identical backpacks for refugees. There are 32 maps and 24 dictionaries to use for the backpacks. What is the greatest number of backpacks they can prepare using all of the maps and dictionaries?

8 backpacks

23. Alyssa is preparing identical fruit baskets. There are 36 oranges and 60 apples to use for the baskets. What is the greatest number of fruit baskets she can prepare using all of the oranges and apples?

12 baskets

LESSON **2-7**

Practice C

Greatest Common Factor

Find the greatest common factor (GCF).

1. 45, 75 — 15
2. 24, 56, 72 — 8
3. 14, 18, 22 — 2
4. 60, 80, 120 — 20
5. 32, 64, 80 — 16
6. 32, 40, 120 — 8
7. 55, 115 — 5
8. 64, 96 — 32
9. 14, 35, 56, 77 — 7

10. Which pair of numbers has a GCF that is a prime number, 91 and 117 or 84 and 156?

91 and 117 have a GCF of 13.

11. Which pair of numbers has a GCF that is a prime number, 54 and 99 or 80 and 114?

80 and 114 have a GCF of 2.

12. Which pair of numbers has a GCF that is a prime number, 70 and 112 or 68 and 187?

68 and 187 have a GCF of 17.

13. Carvin is preparing identical breakfast gift sets. There are 18 boxes of pancake mix and 27 jars of jam. What is the greatest number of gift sets he can prepare using all of the pancake mix and jam?

9 gift sets

14. Amelia is decorating identical frames. She has 75 red feathers and 30 blue tiles. What is the greatest number of frames she can decorate using all of the feathers and tiles?

15 frames

15. A jeweler is making identical necklaces. He has 24 gold beads and 144 colored glass beads. What is the greatest number of necklaces he can make using all the gold beads and glass beads?

24 necklaces

LESSON **2-7**

Reteach

Greatest Common Factor

The **greatest common factor** (GCF) of a group of numbers is the greatest number that can divide each of the numbers without having a remainder.

Use a list to find the GCF of 45, 60, and 75.
List all the factors of each number.

45:	1,		3,		5,	9,			15,			45		
60:	1,	2,	3,	4,	5,	6,	10,	12,	15,	20,	30,		60	
75:	1,		3,		5,				15,	25,				75

The common factors are 1, 3, 5, and 15.
The GCF is 15.

List the factors of each number. Find the greatest common factor (GCF).

1. 9: 1, 3, 9
   24: 1, 2, 3, 4, 6, 8, 12, 24
   GCF = 3
2. 18: 1, 2, 3, 6, 9, 18
   54: 1, 2, 3, 6, 9, 18, 27, 54
   GCF = 18
3. 45: 1, 3, 5, 9, 15, 45
   55: 1, 5, 11, 55
   GCF = 5
4. 12: 1, 2, 3, 4, 6, 12
   20: 1, 2, 4, 5, 10, 20
   GCF = 4
5. 6: 1, 2, 3, 6
   12: 1, 2, 3, 4, 6, 12
   24: 1, 2, 3, 4, 6, 8, 12, 24
   GCF = 6
6. 6: 1, 2, 3, 6
   18: 1, 2, 3, 6, 9, 18
   27: 1, 3, 9, 27
   GCF = 3

Find the greatest common factor (GCF).

7. 27, 45 — 9
8. 14, 18 — 2
9. 12, 30 — 6
10. 10, 45, 70 — 5

LESSON **2-7**

Challenge

The School Trip Factor

There are 24 students in the seventh grade and 36 students in the eighth grade. Solve each problem below about their school trip. Then use your solutions to complete the permission slip each student must have signed in order to go on the trip.

1. The students will visit the zoo, the museum, or the planetarium. Each class will buy its own tickets. Both classes must agree upon the trip that will cost the least. Use the table at right to decide which trip the students will choose.

Trip	Single Tickets	Group Rate	Number of Tickets per Group
Zoo	$8	$24	6
Museum	$5	$12	4
Planetarium	$4	$30	12

2. The principal will provide transportation for both classes. The school PTA will pay the cost but wants the most cost-efficient method of transportation. Use the table at right to decide which method of transportation the PTA will choose.

Type of Transportation	Number of Passengers	Cost
Mini-bus	12	$35
School Bus	70	$185
Mini-van	10	$30

3. The class student council will provide lunch for each class. Each class must agree on the lowest price and then buy at least one lunch pack per student. Use the table at right to decide which lunch deal the students will choose.

Lunch	Number of Lunches per Pack	Cost per Pack
Sandwich and Juice Packs	6	$15
Yogurt and Drink Packs	8	$20
Soup and Sandwich Packs	4	$12

To the Seventh and Eighth-grade Parents:
On Friday, your child will be taking a school trip to the planetarium. The school will pay $60 for the seventh-graders' tickets and $90 for the eigth-graders' tickets. The PTA has graciously offered to pay $175 for 5 [number] mini-buses [vehicles] to transport the students. Each class's student council will pay for its class's lunches. Both classes will buy 10 [number] sandwich and juice [type] packs at a total cost of $60 for the seventh graders and $90 for the eighth graders.

________________ Parent's signature for student's permission

LESSON **2-7**

Problem Solving

Greatest Common Factor

Write the correct answer.

1. Fabric is sold in stores from bolts that are 45 or 60 inches wide. What is the width of the widest strips of fabric you can cut from either bolt without wasting any of the fabric if each strip has the same width?
   15 inches
2. The parents are making sandwiches for the class picnic. They have 72 turkey slices, 48 cheese slices, and 96 tomato slices. What is the greatest number of sandwiches they can make if each sandwich has the same filling?
   24 sandwiches
3. Two bicycle enthusiasts are leaving Cincinnati at the same time. One is biking 840 miles to Boston. The other is biking 440 miles to Atlanta. What is the greatest number of miles a day each can bike if they want to cover equal distances each day?
   40 miles per day
4. A fruit salad made on a TV cooking program requires chunks of cantaloupe and honeydew. What is the greatest number of servings you can make using all of the fruit if you have 30 chunks of cantaloupe and 42 chunks of honeydew?
   6 servings

Choose the letter for the best answer.

5. Ari is making patriotic pins. He has 105 red ribbons, 147 white ribbons, and 189 blue ribbons. What is the greatest number of identical pins he can make if he uses all his ribbons?
   A 52 pins **C** 93 pins
   B 19 pins **(D)** 21 pins
6. Cheryce is making fruit baskets. She has 60 bananas, 72 pears, 96 apples, and 108 oranges. What is the greatest number of equal baskets she can make with the fruit?
   F 3 baskets **(H)** 12 baskets
   G 6 baskets **J** 24 baskets
7. There are 100 senators and 435 representatives in the United States Congress. How many identical groups could be formed from all the senators and representatives?
   A 1 group
   (B) 5 groups
   C 10 groups
   D 15 groups
8. There are 14 baseball teams in the American League and 16 teams in the National League. There are 30 National Hockey League teams. If equal groups of teams are formed, how many hockey teams will be in each group?
   F 3 hockey teams
   G 8 hockey teams
   H 7 hockey teams
   (J) 15 hockey teams

LESSON **2-7**

Reading Strategies

Follow a Procedure

The **greatest common factor** of two numbers is the largest factor the numbers share or have in common. To find the greatest common factor for 36 and 45, follow these steps.

Step 1: Make a list of the factors for each number.

36	45
1 • 36	1 • 45
2 • 18	3 • 15
3 • 12	5 • 9
4 • 9	
6 • 6	

→ **Step 2:** Put the factors in order from smallest to largest.
36: **1**, 2, **3**, 4, 6, **9**, 12, 18, 36
45: **1**, **3**, 5, **9**, 15, 45

Step 3: Find the common factors or both numbers.
1, 3, and **9**

→ **Step 4:** Find the greatest common factor for both numbers.
9

Answer each question.

1. What are the factors for 18? 1, 18, 2, 9, 3, 6
2. Put them in order from least to greatest. 1, 2, 3, 6, 9, 18
3. What are the factors for 21? 1, 21, 7, 3
4. Put them in order from least to greatest. 1, 3, 7, 21
5. Compare your answers for Exercise 2 and Exercise 4. What is the greatest common factor? 3
6. Make a list of all the factors for 48.

48
1 • 48
2 • 24
3 • 16
4 • 12
6 • 8

7. Make a list of all the factors for 60.

60
1 • 60
2 • 30
3 • 20
4 • 15
5 • 12
6 • 10

8. Compare the lists. What factors do both numbers have in common? 1, 2, 3, 4, 6, 12
9. What is the greatest common factor for 48 and 60? 12

LESSON 2-7

Puzzles, Twisters & Teasers

That's the Greatest!

Find the greatest common factor for each set of numbers. Then use the decoder to answer the riddle below.

1. 24, 36, 48, 60 ___12___ L
2. 30, 45, 60, 75 ___15___ E
3. 20, 40, 50, 120 ___10___ K
4. 4, 6, 10, 22 ___2___ N
5. 6, 15, 33, 48 ___3___ A
6. 14, 21, 35, 70 ___7___ R
7. 2, 3, 4, 5, 7 ___1___ O
8. 18, 45, 63, 81 ___9___ C

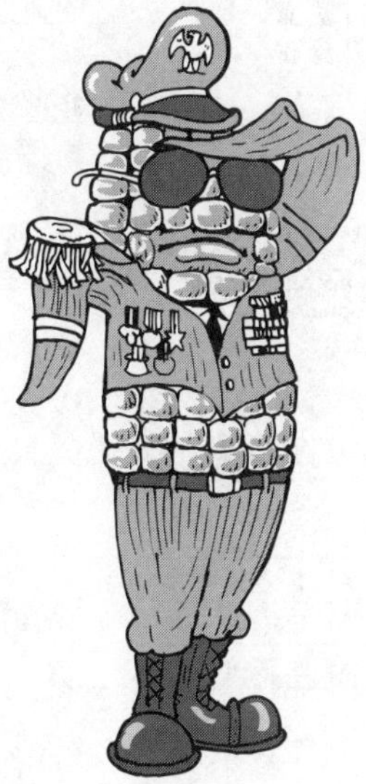

What do you call a piece of corn in the army?

A		C	O	R	N
3		9	1	7	2

K	E	R	N	E	L
10	15	7	2	15	12

LESSON 2-8

Practice A

Least Common Multiple

Find the correct least common multiple (LCM) for each set of numbers below. Match each LCM with a number in the box.

21 60 33 20 35 75
42 20 40 78 99 36

1. 3, 7 ___21___
2. 4, 5 ___20___
3. 10, 12 ___60___
4. 15, 25 ___75___
5. 3, 11 ___33___
6. 9, 12, 18 ___36___
7. 6, 7, 21 ___42___
8. 5, 8, 20 ___40___
9. 2, 26, 39 ___78___

Find the LCM.

10. 10, 18, 45 ___90___
11. 5, 6, 20 ___60___
12. 10, 15, 30 ___30___
13. 2, 4, 7, 8 ___56___
14. 3, 9, 12 ___36___
15. 2, 5, 7, 10 ___70___
16. 5, 9, 18 ___90___
17. 4, 10, 12, 15 ___60___
18. 4, 9, 12, 18 ___36___
19. 8, 16, 32 ___32___
20. 5, 15, 20 ___60___
21. 2, 6, 18, 21 ___126___

22. Carrie and Lance are playing the drums for the band. In the song, Carrie beats the bass drum every 3 beats. Lance beats the snare drum every 5 beats. On what beat will they first play their drums together?

    ___15th beat___

23. Domingo is helping his father paint some boards. Domingo paints a board every 12 minutes. His father paints a board every 8 minutes. If they start together, how long before they will finish painting a board at the same time?

    ___24 minutes___

LESSON 2-8

Practice B

Least Common Multiple

Find the least common multiple (LCM).

1. 8, 10 ___40___
2. 10, 15 ___30___
3. 6, 9 ___18___
4. 12, 16 ___48___
5. 18, 30 ___90___
6. 5, 11 ___55___
7. 15, 45 ___45___
8. 7, 28 ___28___
9. 4, 14 ___28___
10. 3, 10, 12 ___60___
11. 9, 36, 60 ___180___
12. 5, 15 ___15___
13. 7, 14, 49 ___98___
14. 8, 12, 24, 96 ___96___
15. 5, 25, 30 ___150___
16. 5, 9, 18 ___90___
17. 4, 10, 12, 15 ___60___
18. 4, 9, 12, 18 ___36___
19. 4, 12, 24, 36 ___72___
20. 24, 30, 48, 60 ___240___
21. 5, 9, 15, 18 ___90___

22. Jasmine is helping her father plant trees to create a border around the back yard. Jasmine plants a tree every 25 minutes, and her father plants a tree every 15 minutes. If they started together, how long before they would finish planting a tree at the same time?

    ___75 minutes___

23. Two dancers are rehearsing in a studio. One dancer's routine lasts 12 minutes. The other dancer's routine lasts 15 minutes. If they start together and take no breaks between their routines, how long before they start together again?

    ___60 minutes___

24. Evan and Renzo are swimming laps in the pool. It takes Evan 8 minutes to complete 1 lap and Renzo 6 minutes to complete 1 lap. They start together at the tops of their lanes. In how many minutes will they be together again at the tops of their lanes?

    ___24 minutes___

LESSON 2-8

Practice C

Least Common Multiple

Find the least common multiple (LCM).

1. 9, 12 ___36___
2. 18, 60 ___180___
3. 9, 15 ___45___
4. 20, 100 ___100___
5. 7, 11 ___77___
6. 8, 12, 16 ___48___
7. 3, 7, 21 ___21___
8. 40, 100 ___200___
9. 5, 9, 18 ___90___
10. 2, 26, 39 ___78___
11. 5, 8, 12 ___120___
12. 20, 40, 50 ___200___
13. 8, 9, 12 ___72___
14. 4, 6, 9, 12 ___36___
15. 3, 9, 12, 72 ___72___
16. 4, 10, 12 ___60___
17. 5, 20, 25, 50 ___100___
18. 4, 7, 8, 28 ___56___
19. 15, 20, 30 ___60___
20. 2, 6, 9, 21 ___126___
21. 10, 32, 64, 80 ___320___

22. Josh and Eva are packing first-aid supplies. It takes Josh 12 minutes to fill a box with bandages. It takes Eva 8 minutes to fill a box with antibiotics. If they start packing boxes at the same time, how long will it be before they start filling a new box at the same time?

    ___24 minutes___

23. Sadie is helping her mother stain some shelves. Sadie stains a shelf every 25 minutes. Her mother stains a shelf every 15 minutes. If they start together, how long will it be before they finish staining a shelf at the same time?

    ___75 minutes___

24. It takes Amanda 14 minutes to climb up the hill and sled down. It takes Alvin 12 minutes to climb up the hill and sled down. If they start up the hill at the same time, how long will it be before they meet again at the starting point?

    ___84 minutes___

LESSON 2-8 **Reteach**

Least Common Multiple

The **least common multiple** (LCM) of two numbers is the least multiple that the two numbers have in common.

- To list the multiples of a number, multiply the number by 1, 2, 3, and so on.

	8 • 1	8 • 2	8 • 3	8 • 4	8 • 5
Multiples of 8	8	16	24	32	40

	6 • 1	6 • 2	6 • 3	6 • 4	6 • 5
Multiples of 6	6	12	18	24	30

- To find the least common multiple of 6 and 8, list several multiples of each number.

Step 1: List multiples of the greater number. — Multiples of 8: 8 16 **24** 32 40

Step 2: List the multiples of the lesser number until you find a multiple that is common to each list. — Multiples of 6: 6 12 18 **24**

Since 24 is the first number common to both lists, it is the least common multiple. So the LCM of 6 and 8 is 24.

Find the least common multiple (LCM).

1. 4, 9
   Multiples of 9: 9, 18, 27, 36, 45, 54
   Multiples of 4: 4, 8, 12, 16, 20, 24, 28, 32, 36
   LCM of 4 and 9: 36
2. 6, 7
   Multiples of 7: 7, 14, 21, 28, 35, 42, 49, 56
   Multiples of 6: 6, 12, 18, 24, 30, 36, 42
   LCM of 6 and 7: 42
3. 15, 20
   Multiples of 20: 20, 40, 60, 80, 100
   Multiples of 15: 15, 30, 45, 60
   LCM of 15 and 20: 60

4. 12, 18 36 5. 16, 20 80 6. 4, 6, 9 36

LESSON 2-8 **Challenge**

Multiple Solutions

Write the two numbers from each box that have the lowest least common multiple. Then write the LCM.

1. Box: 5 12 / 6 8 → Numbers: 6; 12 LCM: 12
2. Box: 2 7 / 9 10 → Numbers: 2; 10 LCM: 10
3. Box: 2 10 / 5 3 → Numbers: 2; 3 LCM: 6
4. Box: 7 2 / 10 12 → Numbers: 2; 10 LCM: 10
5. Box: 6 4 / 7 5 → Numbers: 4; 6 LCM: 12
6. Box: 9 10 / 15 45 → Numbers: 10; 15 LCM: 30
7. Box: 8 10 / 6 5 → Numbers: 5; 10 LCM: 10
8. Box: 24 35 / 32 14 → Numbers: 14; 35 LCM: 70
9. Box: 15 10 / 7 8 → Numbers: 10; 15 LCM: 30
10. Box: 3 10 / 12 7 → Numbers: 3; 12 LCM: 12

Use the table at the bottom of the page to solve the riddle. In the middle row, write the LCM of each exercise above the exercise number. Then choose the letter from the Answer Key that matches the LCM and write it in the appropriate box.

Answer Key

C	A	T	E	H
70	6	30	12	10

11. What animal copies from the others on a math test?

T	H	E		C	H	E	E	T	A	H
30	10	12		70	10	12	12	30	6	10
9	4	5		8	2	1	5	6	3	7

LESSON 2-8 **Problem Solving**

Least Common Multiple

Write the correct answer.

1. Earth revolves around the sun every year. Jupiter revolves around the sun every 12 years. If Earth and Jupiter passed the same point of the sun sometime in 2002, when will they pass that point together again?
   2014
2. House representatives are elected every 2 years. The President of the United States is elected every 4 years. Both will be elected in 2008. When is the next year after 2008 both will be elected?
   2012
3. A cat runs a mile every 2 minutes. A squirrel runs a mile every 5 minutes. A cat and a squirrel start together running around a 1-mile track. How long will it be before they meet at the starting point?
   10 minutes
4. A car manual recommends changing the oil every 5,000 miles and inspecting the engine coolant system every 15,000 miles. At how many miles will both be done together for the first time? for the second time?
   15,000 mi; 30,000 mi

Choose the letter for the best answer.

5. Mr. Walters receives a dividend every 5 months and a royalty payment every 6 months. He received both in January 2006. When is the next time he would receive both payments in the same month?
   A January 2007
   B April 2008
   C July 2006
   (D) July 2008
6. Rag Rite Cloth Store always rounds amounts less than whole yards up to the next yard for ribbon purchases. The ribbon that you want to buy comes in rolls of 8 feet. How many rolls should you buy to get the best buy?
   F 1 roll
   G 2 rolls
   (H) 3 rolls
   J 4 rolls
7. The sanitation department picks up recyclable plastics every 3 days. The paper recycling center picks up papers every 4 days. They both picked up on May 4. When will they next pick up on the same day?
   A May 7
   B May 12
   C May 11
   (D) May 16
8. Hal and Jess both volunteer at the local nursing home. Hal volunteers every 6 days, and Jess volunteers every 8 days. They were both there on Monday. In how many days will they both volunteer together again?
   F in 14 days
   (G) in 24 days
   H in 48 days
   J in 2 days

LESSON 2-8 **Reading Strategies**

Focus on Vocabulary

A **multiple** of a number is the product of that number and another factor.

Multiples of 5 are the numbers you get when you multiply by **5.**

1 • 5 = **5**   2 • 5 = **10**   3 • 5 = **15**   4 • 5 = **20**

The multiples of 5 listed above are 5, 10, 15, 20.

You can continue to count by 5's to list more multiples of 5: 25, 30, 35, and so on.

Common means something that is shared. A common multiple of two numbers is a multiple that the two numbers share, or have in common.

Multiples of 5 → 5, 10, 15, **20,** 25, 30, 35, **40,** 45, 50, 55, **60**

Multiples of 4 → 4, 8, 12, 16, **20,** 24, 28, 32, 36, **40,** 44, 48, 52, 56, **60**

In the example above, you can see the multiples that are common to both 4 and 5. They are 20, 40, and 60.

The **least common multiple (LCM)** of two numbers is the smallest of the common multiples. 20 is the smallest, or least common, multiple of 4 and 5.

Answer each question.

1. What is a multiple?
   Possible answer: the answers or products when you multiply by a certain number
2. What does the word *common* mean?
   something that is shared
3. What is a common multiple?
   multiples that two numbers share, or have in common
4. What is the least common multiple?
   the smallest multiple that two numbers have in common
5. Write the first 9 multiples of 2. 2, 4, 6, 8, 10, 12, 14, 16, 18
6. Write the first 9 multiples of 3. 3, 6, 9, 12, 15, 18, 21, 24, 27
7. Write the common multiples for both 2 and 3. 6, 12, and 18
8. What is the least common multiple for 2 and 3? 6

LESSON **2-8** **Puzzles, Twisters & Teasers**

A Common Cause!

Find the least common multiple for each set of numbers. Spell out the answers to solve the crossword puzzle.

Across	Down
1. 3, 5	**1.** 3, 9, 15
4. 9, 12	**2.** 4, 6, 24
6. 10, 4	**3.** 6, 9
7. 2, 5	**5.** 4, 7
9. 6, 14	**6.** 3, 7
	8. 15, 20

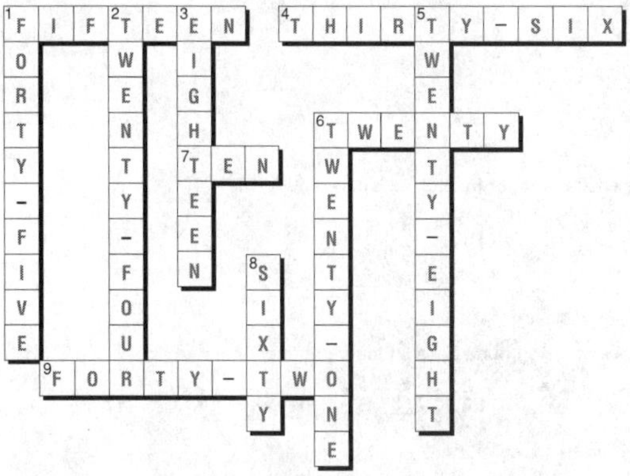

LESSON **2-9** **Practice A**

Equivalent Fractions and Mixed Numbers

Find a fraction equivalent to each fraction. Possible answers given.

1. $\frac{1}{4}$ $\frac{2}{8}$ **2.** $\frac{3}{5}$ $\frac{6}{10}$ **3.** $\frac{4}{9}$ $\frac{8}{18}$ **4.** $\frac{1}{8}$ $\frac{2}{16}$

5. $\frac{1}{6}$ $\frac{2}{12}$ **6.** $\frac{20}{40}$ $\frac{1}{2}$ **7.** $\frac{6}{15}$ $\frac{2}{5}$ **8.** $\frac{35}{50}$ $\frac{7}{10}$

Write the fractions with a common denominator. Then determine if they are equivalent. Possible answers given.

9. $\frac{2}{3}$ and $\frac{6}{9}$ — $\frac{6}{9}, \frac{6}{9}$; yes **10.** $\frac{7}{10}$ and $\frac{8}{15}$ — $\frac{21}{30}, \frac{16}{30}$; no **11.** $\frac{3}{6}$ and $\frac{4}{8}$ — $\frac{12}{24}, \frac{12}{24}$; yes

12. $\frac{16}{8}$ and $\frac{10}{5}$ — $\frac{80}{40}, \frac{80}{40}$; yes **13.** $\frac{9}{6}$ and $\frac{6}{4}$ — $\frac{18}{12}, \frac{18}{12}$; yes **14.** $\frac{8}{12}$ and $\frac{15}{24}$ — $\frac{16}{24}, \frac{15}{24}$; no

Write each as a mixed number.

15. $\frac{5}{4}$ $1\frac{1}{4}$ **16.** $\frac{9}{2}$ $4\frac{1}{2}$ **17.** $\frac{13}{5}$ $2\frac{3}{5}$ **18.** $\frac{12}{7}$ $1\frac{5}{7}$

Write each as an improper fraction.

19. $3\frac{2}{3}$ $\frac{11}{3}$ **20.** $2\frac{3}{8}$ $\frac{19}{8}$ **21.** $1\frac{3}{5}$ $\frac{8}{5}$ **22.** $2\frac{5}{6}$ $\frac{17}{6}$

23. Lou's bedroom is $11\frac{1}{2}$ feet long. Write this number as an improper fraction.

$\frac{23}{2}$

24. Kaleigh used $\frac{2}{3}$ cup of flour to make cookies. Eric used $\frac{9}{12}$ cup of flour. Did they use the same amount of flour?

no

LESSON **2-9** **Practice B**

Equivalent Fractions and Mixed Numbers

Find a fraction equivalent to the given fraction. Possible answers given.

1. $\frac{2}{9}$ $\frac{4}{18}$ **2.** $\frac{8}{15}$ $\frac{16}{30}$ **3.** $\frac{7}{8}$ $\frac{14}{16}$

4. $\frac{16}{24}$ $\frac{2}{3}$ **5.** $\frac{12}{20}$ $\frac{3}{5}$ **6.** $\frac{9}{12}$ $\frac{3}{4}$

Write the fractions with a common denominator. Then determine if they are equivalent. Possible answers given.

7. $\frac{8}{10}$ and $\frac{12}{15}$ — $\frac{24}{30}, \frac{24}{30}$; yes **8.** $\frac{6}{8}$ and $\frac{8}{12}$ — $\frac{18}{24}, \frac{16}{24}$; no **9.** $\frac{3}{9}$ and $\frac{4}{8}$ — $\frac{24}{72}, \frac{36}{72}$; no

10. $\frac{7}{4}$ and $\frac{9}{5}$ — $\frac{35}{20}, \frac{36}{20}$; no **11.** $\frac{15}{12}$ and $\frac{20}{16}$ — $\frac{60}{48}, \frac{60}{48}$; yes **12.** $\frac{15}{9}$ and $\frac{30}{18}$ — $\frac{30}{18}, \frac{30}{18}$; yes

Write each as a mixed number.

13. $\frac{21}{8}$ $2\frac{5}{8}$ **14.** $\frac{37}{4}$ $9\frac{1}{4}$ **15.** $\frac{16}{5}$ $3\frac{1}{5}$ **16.** $\frac{49}{9}$ $5\frac{4}{9}$

Write each as an improper fraction.

17. $8\frac{2}{3}$ $\frac{26}{3}$ **18.** $1\frac{7}{12}$ $\frac{19}{12}$ **19.** $25\frac{3}{4}$ $\frac{103}{4}$ **20.** $7\frac{5}{6}$ $\frac{47}{6}$

21. Maria's desk is $33\frac{3}{4}$ inches long. Write this number as an improper fraction.

$\frac{135}{4}$

22. Leon walked $\frac{5}{8}$ mile. Liz walked $\frac{10}{16}$ mile. Did they walk the same distance?

yes

LESSON **2-9** **Practice C**

Equivalent Fractions and Mixed Numbers

Find two fractions equivalent to each number. Possible answers given.

1. $\frac{7}{12}$ $\frac{14}{24}, \frac{21}{36}$ **2.** $\frac{11}{15}$ $\frac{22}{30}, \frac{33}{45}$ **3.** $\frac{14}{25}$ $\frac{28}{50}, \frac{42}{75}$

4. $\frac{18}{20}$ $\frac{9}{10}, \frac{36}{40}$ **5.** 5 $\frac{5}{1}, \frac{10}{2}$ **6.** $1\frac{1}{8}$ $\frac{9}{8}, \frac{18}{16}$

Write the fractions with a common denominator. Then determine if they are equivalent. Possible answers given.

7. $\frac{10}{15}$ and $\frac{12}{20}$ — $\frac{40}{60}, \frac{36}{60}$; no **8.** $\frac{10}{18}$ and $\frac{9}{12}$ — $\frac{20}{36}, \frac{27}{36}$; no **9.** $\frac{18}{24}$ and $\frac{12}{16}$ — $\frac{36}{48}, \frac{36}{48}$; yes

10. $\frac{7}{8}$ and $\frac{9}{10}$ — $\frac{35}{40}, \frac{36}{40}$; no **11.** $\frac{32}{6}$ and $\frac{48}{9}$ — $\frac{96}{18}, \frac{96}{18}$; yes **12.** $\frac{38}{15}$ and $\frac{25}{10}$ — $\frac{76}{30}, \frac{75}{30}$; no

Write each as a mixed number.

13. $\frac{43}{7}$ $6\frac{1}{7}$ **14.** $\frac{27}{5}$ $5\frac{2}{5}$ **15.** $\frac{55}{8}$ $6\frac{7}{8}$ **16.** $\frac{80}{11}$ $7\frac{3}{11}$

Write each as an improper fraction.

17. $25\frac{2}{5}$ $\frac{127}{5}$ **18.** $11\frac{7}{9}$ $\frac{106}{9}$ **19.** $4\frac{11}{12}$ $\frac{59}{12}$ **20.** $5\frac{5}{16}$ $\frac{85}{16}$

21. Write an improper fraction and a mixed number to represent 139 inches in feet.

$\frac{139}{12}$, $11\frac{7}{12}$

22. A baker used $52\frac{1}{2}$ cups of flour to make 15 loaves of bread. How many quarter cups of flour were used?

210 quarter cups

LESSON **2-9** **Reteach**

Equivalent Fractions and Mixed Numbers

Equivalent fractions name the same number.

- You can use fraction strips to find equivalent fractions.

Find two fractions equivalent to $\frac{6}{8}$.

$\frac{1}{8}$	$\frac{1}{8}$	$\frac{1}{8}$	$\frac{1}{8}$	$\frac{1}{8}$	$\frac{1}{8}$

First, use fraction strips to model $\frac{6}{8}$.

$\frac{1}{8}$	$\frac{1}{8}$	$\frac{1}{8}$	$\frac{1}{8}$	$\frac{1}{8}$	$\frac{1}{8}$	$\frac{6}{8}$

Then find equivalent fractions by placing other fraction strips with the same length below the model.

$\frac{1}{4}$	$\frac{1}{4}$	$\frac{1}{4}$	$\frac{3}{4}$

$\frac{3}{4}$ and $\frac{12}{16}$ are equivalent to $\frac{6}{8}$.

$\frac{1}{16}$	$\frac{1}{16}$	$\frac{1}{16}$	$\frac{1}{16}$	$\frac{1}{16}$	$\frac{1}{16}$	$\frac{1}{16}$	$\frac{1}{16}$	$\frac{1}{16}$	$\frac{1}{16}$	$\frac{1}{16}$	$\frac{1}{16}$	$\frac{12}{16}$

Use fractions strips to find a fraction equivalent to each fraction. Possible answers given.

1. $\frac{1}{2}$ $\underline{\frac{2}{4}}$ 2. $\frac{1}{4}$ $\underline{\frac{2}{8}}$ 3. $\frac{2}{5}$ $\underline{\frac{4}{10}}$ 4. $\frac{2}{3}$ $\underline{\frac{4}{6}}$

- You can use cross products to determine if two fractions are equivalent.

$\frac{8}{10}$ ✕ $\frac{12}{15}$ $8 \cdot 5 \stackrel{?}{=} 10 \cdot 12$

$120 = 120$

$\frac{8}{10}$ and $\frac{12}{15}$ are equivalent. Their cross products are equal.

$\frac{9}{15}$ ✕ $\frac{2}{3}$ $9 \cdot 3 \stackrel{?}{=} 15 \cdot 2$

$27 \neq 30$

$\frac{9}{15}$ and $\frac{2}{3}$ are not equivalent. Their cross products are not equal.

Find the cross products. Determine if the two fractions are equivalent. Write *yes* or *no*.

5. $\frac{18}{6}$ and $\frac{10}{3}$ 54; 60; no
6. $\frac{6}{10}$ and $\frac{9}{15}$ 90; 90; yes
7. $\frac{2}{3}$ and $\frac{9}{12}$ 24; 27; no

LESSON **2-9** **Reteach**

Equivalent Fractions and Mixed Numbers (continued)

In an *improper fraction* the numerator is greater than the denominator, such as $\frac{11}{3}$.

A *mixed number* is made up of both a whole number part and a fraction part, such as $2\frac{5}{8}$.

- To write an improper fraction as a mixed number, you can write the improper fraction as a sum of fractions.

$\frac{11}{3} = \underbrace{\frac{3}{3} + \frac{3}{3} + \frac{3}{3}}_{1 + 1 + 1} + \frac{2}{3} = 3\frac{2}{3}$

- To write a mixed number as an improper fraction, you can write the whole number as a fraction with the same denominator as the fraction part.

$2\frac{5}{8} = 2 + \frac{5}{8} = \frac{16}{8} + \frac{5}{8} = \frac{21}{8}$

8. Write $\frac{13}{4}$ as a mixed number.

$\frac{13}{4} = \frac{4}{4} + \frac{4}{4} + \frac{4}{4} + \frac{1}{4} = \underline{3\frac{1}{4}}$ (↑ 1 ↑ 1 ↑ 1)

9. Write $3\frac{2}{5}$ as an improper fraction.

$3\frac{2}{5} = \underline{3} + \frac{2}{5} = \frac{15}{5} + \frac{2}{5} = \frac{17}{5}$

Write each improper fraction as a mixed number.

10. $\frac{23}{5}$ $\underline{4\frac{3}{5}}$ 11. $\frac{15}{8}$ $\underline{1\frac{7}{8}}$ 12. $\frac{54}{7}$ $\underline{7\frac{5}{7}}$ 13. $\frac{14}{3}$ $\underline{4\frac{2}{3}}$

Write each mixed number as an improper fraction.

14. $5\frac{1}{2}$ $\underline{\frac{11}{2}}$ 15. $3\frac{3}{8}$ $\underline{\frac{27}{8}}$ 16. $2\frac{11}{15}$ $\underline{\frac{41}{15}}$ 17. $8\frac{3}{4}$ $\underline{\frac{35}{4}}$

LESSON **2-9** **Challenge**

Target Practice

Find out where the dart lands on the dartboard. Cross out the fraction, improper fraction, or mixed number on the dartboard that is equivalent to each number below. The dart lands on the number that is left.

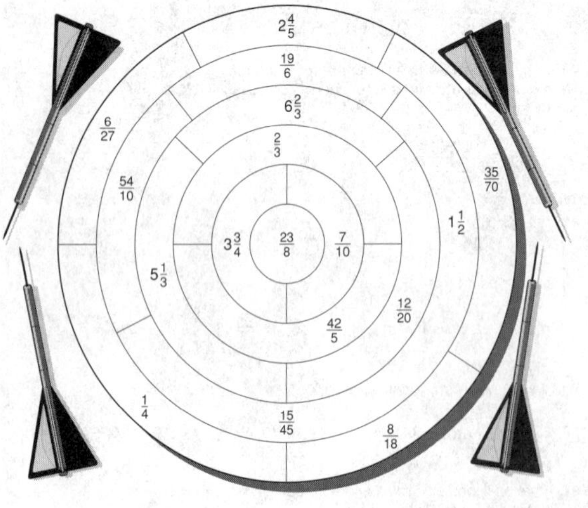

1. $\frac{32}{6}$ 2. $\frac{14}{20}$ 3. $\frac{12}{8}$ 4. $\frac{3}{5}$
5. $2\frac{7}{8}$ 6. $\frac{28}{10}$ 7. $\frac{16}{36}$ 8. $\frac{50}{75}$
9. $\frac{2}{9}$ 10. $\frac{20}{3}$ 11. $\frac{5}{15}$ 12. $3\frac{1}{6}$
13. $\frac{3}{12}$ 14. $5\frac{2}{5}$ 15. $\frac{45}{12}$ 16. $\frac{1}{2}$

17. The dart lands on $\underline{\frac{42}{5}}$

LESSON **2-9** **Problem Solving**

Equivalent Fractions and Mixed Numbers

Write the correct answer.

1. Bert bicycles 136 miles each week. If he bikes the same distance daily and bikes each day of the week, how many miles does he bike each day? Write your answer as a mixed number.

$19\frac{3}{7}$ miles

2. The Eiffel Tower in Paris, France, was designed so well that even in the highest winds the tower never sways more than $4\frac{1}{2}$ inches. Write two fractions that are equivalent to $4\frac{1}{2}$.

Possible answer: $\frac{9}{2}$, $\frac{18}{4}$

3. In October 2000, almost 13 inches of rain fell in Miami, Florida, in a 24-hour period. Write a fraction that shows about how many $\frac{1}{4}$ inches of rain fell on that one day.

$\frac{52}{4}$ inches

4. Elena is 58 inches tall. Write a mixed number to represent her height in feet.

$4\frac{5}{6}$ feet

Choose the letter for the best answer.

5. Randi runs 48 miles each week. She runs the same distance daily and runs each day of the week. How many miles does she run each day? Write your answer as a mixed number.

Ⓐ $6\frac{6}{7}$ miles C $6\frac{7}{48}$ miles

B $6\frac{5}{6}$ miles D $7\frac{1}{7}$ miles

6. A bakery used $4\frac{3}{4}$ cups of salt to make 120 loaves of sourdough bread. Write a fraction that shows how many $\frac{1}{4}$ cups of salt were used to make the bread.

F $\frac{30}{4}$ cups H $\frac{12}{4}$ cups

G $\frac{24}{4}$ cups Ⓙ $\frac{19}{4}$ cups

7. One of the driest summers in the Northeast was in 1909, with $\frac{896}{100}$ inches of precipitation. Which mixed number is equivalent to $\frac{896}{100}$?

A $89\frac{6}{10}$ Ⓒ $8\frac{24}{25}$

B $8\frac{96}{10}$ D $8\frac{4}{5}$

8. The Central Ohio Transit Authority has $957\frac{3}{5}$ miles of routes. Which fraction is equivalent to $957\frac{3}{5}$?

F $\frac{4,781}{5}$ H $\frac{4,783}{5}$

Ⓖ $\frac{4,788}{5}$ J $\frac{9,573}{5}$

LESSON 2-9

Reading Strategies

Multiple Meaning Terms

If you multiply or divide the numerator and denominator of a fraction by the same number, you get another fraction that is **equivalent,** or equal.

$\frac{2}{3} \cdot \frac{2}{2} = \frac{4}{6}$

In the example above, both the numerator and denominator are multiplied by 2. Equal fractions are called **equivalent fractions.**

$\frac{2}{3}$ is equivalent to $\frac{4}{6}$.

Answer the following questions.

1. Is $\frac{1}{2}$ equivalent to $\frac{6}{12}$? yes
2. How do you know?

   Possible answer: You can multiply both the numerator and denominator by 6 to get $\frac{6}{12}$.
3. Are $\frac{1}{5}$ and $\frac{2}{3}$ equivalent fractions? no
4. How do you know?

   Possible answer: You cannot multiply both 1 and 5 by the same number to get 2 and 3.
5. Is $\frac{7}{15}$ equivalent to $\frac{1}{7}$? no
6. How do you know?

   Possible answer: You cannot divide both 7 and 15 by the same number to get 1 and 7.
7. Is $\frac{3}{9}$ equivalent to $\frac{1}{3}$? yes
8. How do you know?

   Possible answer: You can divide both 3 and 9 by 3 to get $\frac{1}{3}$.

LESSON 2-9

Puzzles, Twisters & Teasers

Are You in the Dark?

Fill in the boxes in the grids to show equivalent fractions. Then use the Answer Key to answer the riddle.

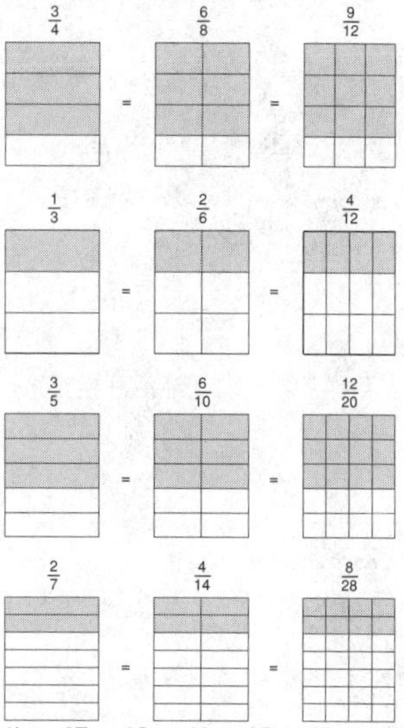

1I 2T 3S 4A 6E 8R 12D

Why did the tomato blush?

I(1) T(2) S(3) A(4) W T(2) H E(6)

S(3) A(4) L A(4) D(12) D(12) R(8) E(6) S(3) S(3) I(1) N G.

LESSON 2-10

Practice A

Equivalent Fractions and Decimals

Write each fraction as a decimal. Round to the nearest hundredth, if necessary.

1. $\frac{7}{10}$ 0.7
2. $\frac{9}{20}$ 0.45
3. $\frac{3}{4}$ 0.75
4. $\frac{20}{25}$ 0.8
5. $\frac{3}{8}$ 0.375
6. $\frac{7}{5}$ 1.4
7. $\frac{21}{7}$ 3.0
8. $\frac{5}{3}$ 1.667
9. $\frac{3}{10}$ 0.3
10. $\frac{4}{5}$ 0.8
11. $\frac{1}{25}$ 0.04
12. $\frac{3}{20}$ 0.15

Write each decimal as a fraction or mixed number in simplest form.

13. 0.55 $\frac{11}{20}$
14. 0.03 $\frac{3}{100}$
15. −0.75 $-\frac{3}{4}$
16. 2.1 $2\frac{1}{10}$ or $\frac{21}{10}$
17. 5.25 $5\frac{1}{4}$ or $\frac{21}{4}$
18. 9.33 $9\frac{33}{100}$ or $\frac{933}{100}$
19. 1.8 $\frac{9}{5}$ or $1\frac{4}{5}$
20. −1.74 $-\frac{87}{50}$ or $-1\frac{37}{50}$
21. 10.6 $\frac{53}{5}$ or $10\frac{3}{5}$
22. −7.08 $-7\frac{2}{25}$ or $-\frac{177}{25}$
23. 0.625 $\frac{5}{8}$
24. 0.001 $\frac{1}{1,000}$

Write each answer as a decimal rounded to the nearest thousandth.

25. Out of 45 times at bat, Raul got 19 hits. Find Raul's batting average.

    0.422
26. On a test, Selena answered 26 out of 30 questions correctly. What portion of her answers was correct?

    0.867

LESSON 2-10

Practice B

Equivalent Fractions and Decimals

Write each fraction as a decimal. Round to the nearest hundredth, if necessary.

1. $\frac{2}{10}$ 0.2
2. $\frac{19}{20}$ 0.95
3. $\frac{5}{8}$ 0.625
4. $\frac{11}{5}$ 2.2
5. $\frac{19}{6}$ 3.167
6. $\frac{17}{4}$ 4.25
7. $\frac{13}{12}$ 1.083
8. $\frac{30}{7}$ 4.286
9. $\frac{7}{4}$ 1.75
10. $\frac{9}{20}$ 0.45
11. $\frac{11}{10}$ 1.1
12. $\frac{2}{25}$ 0.08

Write each decimal as a fraction in simplest form.

13. 0.85 $\frac{17}{20}$
14. 0.11 $\frac{11}{100}$
15. −0.25 $-\frac{1}{4}$
16. 4.3 $4\frac{3}{10}$ or $\frac{43}{10}$
17. 7.75 $7\frac{3}{4}$ or $\frac{31}{4}$
18. 5.03 $5\frac{3}{100}$ or $\frac{503}{100}$
19. −1.06 $-1\frac{3}{50}$ or $-\frac{53}{50}$
20. 0.375 $\frac{3}{8}$
21. −2.65 $-2\frac{13}{20}$ or $-\frac{53}{20}$
22. −5.6 $-5\frac{3}{5}$ or $-\frac{28}{5}$
23. 1.12 $1\frac{3}{25}$ or $\frac{28}{25}$
24. 0.005 $\frac{1}{200}$

Write each answer as a decimal rounded to the nearest thousandth.

25. In the 1998 Winter Olympics, a total of 205 medals were awarded. The United States won 13 medals. What portion of the medals did the United States win?

    0.063
26. On a test, Hailey answered 64 out of 75 questions correctly. What portion of her answers was correct?

    0.853

LESSON **2-10**
Practice C
Equivalent Fractions and Decimals

Write each fraction as a decimal. Round to the nearest hundredth, if necessary.

1. $\frac{5}{12}$ **0.417**
2. $\frac{19}{4}$ **4.75**
3. $\frac{25}{7}$ **3.571**
4. $\frac{9}{8}$ **1.125**
5. $\frac{17}{3}$ **5.667**
6. $\frac{13}{11}$ **1.182**
7. $\frac{15}{9}$ **1.667**
8. $\frac{30}{13}$ **2.308**
9. $\frac{9}{50}$ **0.18**
10. $\frac{13}{25}$ **0.52**
11. $\frac{23}{20}$ **1.15**
12. $\frac{17}{4}$ **4.25**

Write each decimal as a fraction in simplest form.

13. 0.25 **$\frac{1}{4}$**
14. 0.13 **$\frac{13}{100}$**
15. −0.65 **$-\frac{13}{20}$**
16. 4.8 **$4\frac{4}{5}$ or $\frac{24}{5}$**
17. 1.875 **$1\frac{7}{8}$ or $\frac{15}{8}$**
18. 9.06 **$9\frac{3}{50}$ or $\frac{453}{50}$**
19. −1.008 **$-1\frac{1}{125}$ or $-\frac{126}{125}$**
20. 0.625 **$\frac{5}{8}$**
21. −2.05 **$-2\frac{1}{20}$ or $-\frac{41}{20}$**

Determine if the numbers in each pair are equivalent.

22. $\frac{3}{7}$ and 0.41 **no**
23. $\frac{13}{20}$ and 0.65 **yes**
24. $\frac{19}{25}$ and 0.78 **no**

25. In the 1996 Summer Olympics, a total of 842 medals were awarded. The United States won 101 medals. What portion of the medals did the United States win? Write your answer as a decimal rounded to the nearest thousandth.

0.120

26. Pete Rose had 4,256 hits in 14,053 at bats. Write his batting average as a fraction and as a decimal rounded to the nearest thousandth.

$\frac{4,256}{14,053}$, 0.303

LESSON **2-10**
Reteach
Equivalent Fractions and Decimals

To write a fraction as a decimal, divide the numerator of the fraction by the denominator of the fraction.

Write $\frac{3}{7}$ as a decimal.
- Divide 3 by 7.
- To round your answer to the nearest hundredth, add 3 zeros after the decimal point in the divisor.

```
   0.428
7)3.000
 −28↓
   20↓
  −14
    60↓
   −56
```

0.428 rounded to the nearest hundredth is 0.43.

1. Write $\frac{2}{5}$ as a decimal.

```
   0.4
5)2.0
 −2 0
    0
```

$\frac{2}{5}$ = **0.4**

Write each fraction as a decimal. Round to the nearest hundredth, if necessary.

2. $\frac{3}{4}$ **0.75**
3. $\frac{7}{8}$ **0.88**
4. $\frac{3}{2}$ **1.5**
5. $\frac{5}{3}$ **1.67**

To write a decimal as a fraction:
Step 1: Use place value to read the decimal. Say the number aloud.
Step 2: Write a fraction for the number you just said.
Step 3: Simplify if necessary.

Write 0.005 as a fraction.
Read 0.005 as "five thousandths."
Write $\frac{5}{1000}$ for five thousandths.
Simplify: $\frac{5 \div 5}{1{,}000 \div 5} = \frac{1}{200}$

Write 1.6 as a fraction.
Read 1.6 as "one and six tenths."
Write $1\frac{6}{10}$ for one and six tenths.
Simplify: $1\frac{6 \div 2}{10 \div 2} = 1\frac{3}{5}$

Write each decimal as a fraction or mixed number in simplest form.

6. 0.8 **$\frac{4}{5}$**
7. 2.25 **$\frac{9}{4}$ or $2\frac{1}{4}$**
8. −0.02 **$-\frac{1}{50}$**

LESSON **2-10**
Challenge
Decimal Patterns

Repeating decimals continue without end with a repeating pattern. Some decimals continue without end, but do not have a repeating pattern. These decimals are called *nonrepeating, nonterminating* decimals. Although they continue without end, these decimals may have a pattern.

Examples of nonrepeating, nonterminating decimals with patterns:

5.121121112...
0.881882883...

Write *r* if the decimal is a repeating decimal and *n* if the decimal is a nonrepeating, nonterminating decimal.

1. 4.562222... **r**
2. 6.1323323332... **n**
3. 8.355355... **r**
4. 0.230230023... **n**
5. 10.7727227... **n**
6. 4.121314... **n**
7. 0.42818181... **r**
8. 29.7053053... **r**
9. 1.221221222... **n**
10. 5.67167672... **n**
11. 75.320333... **r**
12. 4.121231234... **n**

Identify the pattern in each nonrepeating, nonterminating decimal. Use the pattern to extend the decimal several places.

13. 0.010203... **0.01020304...**
14. 8.636336333... **8.63633633363333...**
15. 25.121231234... **25.12123123412345...**
16. 9.989796... **9.98979695...**

17. Write an example of a repeating decimal.

Possible answer: 1.2222...

18. Write an example of a nonrepeating, nonterminating decimal with a pattern.

Possible answer: 1.191991999...

LESSON **2-10**
Problem Solving
Equivalent Fractions and Decimals

Write the correct answer.

1. On a test, Shane answered 37 out of 40 questions correctly. What portion of his answers was correct? Write your answer as a decimal rounded to the nearest thousandth.

0.925

2. Sammy Sosa hit 66 home runs in 1998. He had 643 at bats. Write his home run average as a decimal rounded to the nearest thousandth.

0.103

3. In February, Chicago receives an average of $1\frac{2}{5}$ inches of rain. Write a decimal to show the number of inches of rain.

1.4 in.

4. On a test, Ellen answered 51 out of 64 questions correctly. What portion of her answers was correct? Write your answer as a decimal rounded to the nearest thousandth.

0.797

Choose the letter for the best answer.

Use the graph for 5–6.

World Energy Consumption 2003
(in quadrillions of British Thermal Units, BTU)

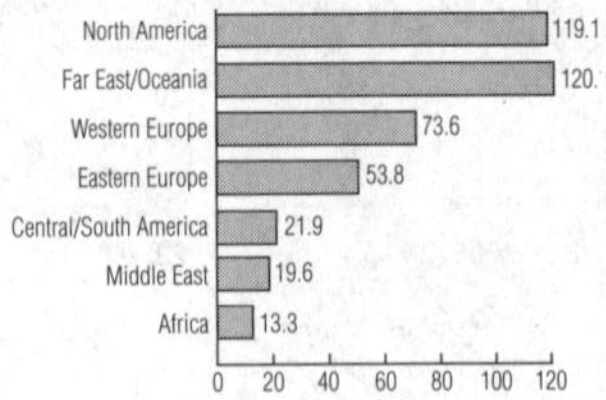

5. Which mixed number shows the energy, in quadrillions of BTU, consumed in Eastern Europe in 2003?

A 53 C $53\frac{1}{2}$
B $53\frac{1}{4}$ **(D)** $53\frac{4}{5}$

6. Which mixed number shows the energy, in quadrillions of BTU, consumed in the Middle East in 2003?

F $19\frac{1}{10}$ **(H)** $19\frac{3}{5}$
G $19\frac{2}{5}$ J $19\frac{3}{10}$

7. Jill sold 478 out of 520 tickets to the opening night of her theater performance. What portion of the tickets did she sell?

A 0.998 **(C)** 0.919
B 1.088 D 0.081

8. The high school sold 369 out of 460 tickets to the opening night of a concert series. What portion of the tickets was sold?

F 0.198 H 1.247
G 0.829 **(J)** 0.802

LESSON **2-10**

Reading Strategies

Compare and Contrast

Compare what happens when fractions are changed to decimals.

$\frac{2}{5}$ • Read $\frac{2}{5}$ as "2 divided by 5." • Write → $2 \div 5$

Change a fraction to a decimal by dividing the numerator by the denominator.

$$5\overline{)2.0} = .4, \quad -2\,0$$

$\frac{2}{5} = 0.4$ The dividing ends, or terminates, with no remainder. 0.4 is called a terminating decimal.

1. Is there a remainder in the problem? How do you know?

   No. Possible answer: When you subtract 20 from 20 there is a remainder of 0.

2. What do we call a decimal that ends with no remainder?

   a terminating decimal

$\frac{2}{6}$ • Read $\frac{2}{6}$ as "2 divided by 6." • Write → $2 \div 6$

$$6\overline{)2.000} = .333$$
$-1\,8$
20
-18
20 ← Note how dividing continues in a pattern. The number 0.333 . . . is a repeating decimal. The bar over the 3 means 3 repeats.
-18
2

$\frac{2}{6} = 0.333\ldots$ or $0.\overline{3}$

Answer each question.

3. Compare the division of $\frac{2}{5}$ to the division of $\frac{2}{6}$. What is the difference?

   Possible answer: There is a remainder when dividing 2 by 6, but no remainder when dividing 2 by 5.

4. What is the name for a decimal with a remainder that has a repeating pattern?

   repeating decimal

LESSON **2-10**

Puzzles, Twisters & Teasers

The Terminator!

Find the words on the list in the word search. Then find four letters in the word search that answer the riddle.

fraction decimal terminating repeating average
statistic convert numerator divide pattern denominator

D	E	C	I	M	A	L	M	H	T	Y	U	J	K	S	T
I	E	R	E	P	E	A	T	I	N	G	C	F	T	T	E
V	W	N	W	E	R	T	Y	U	I	I	C	D	K	A	R
I	F	V	O	Z	X	C	V	B	N	M	K	U	W	T	M
D	R	Q	E	M	T	G	B	Y	H	N	J	L	Z	I	I
E	A	S	D	U	I	B	G	R	V	F	E	C	I	S	N
P	C	O	L	C	O	N	V	E	R	T	P	O	D	T	A
O	T	M	L	K	I	P	A	V	E	R	A	G	E	I	T
I	I	W	E	R	P	A	T	T	E	R	N	T	R	C	I
U	O	Z	A	Q	W	S	X	C	O	T	G	H	J	I	N
Y	N	G	H	J	K	N	U	M	E	R	A	T	O	R	G

What will the dentist say to you if you do not brush your teeth?

I C D K (I see decay.)

LESSON **2-11**

Practice A

Comparing and Ordering Rational Numbers

Compare the fractions. Write < or >.

1. $\frac{4}{9}$ [<] $\frac{7}{9}$
2. $-\frac{5}{6}$ [<] $-\frac{1}{6}$
3. $-\frac{2}{3}$ [>] $-\frac{3}{4}$
4. $\frac{12}{5}$ [>] $1\frac{1}{3}$
5. $\frac{4}{7}$ [>] $-\frac{6}{7}$
6. $-\frac{1}{4}$ [<] $-\frac{1}{7}$
7. $1\frac{5}{8}$ [<] $1\frac{9}{10}$
8. $3\frac{1}{5}$ [>] $-3\frac{2}{5}$

Compare the decimals. Write < or >.

9. 7.3 [<] 7.6
10. 0.053 [<] 0.312
11. −2.51 [>] −2.511
12. 1.450 [>] 1.405
13. 0.445 [<] 0.455
14. $1.\overline{18}$ [<] 1.188
15. −5.21 [<] 5.00
16. $9.\overline{03}$ [<] 9.303

Write the numbers in order from least to greatest.

17. $0.\overline{5}, 0.5, \frac{5}{8}$ — $0.5, 0.\overline{5}, \frac{5}{8}$
18. $1.34, 1\frac{1}{3}, 1.3$ — $1.3, 1\frac{1}{3}, 1.34$
19. $2.07, 2\frac{7}{10}, 2.67$ — $2.07, 2.67, 2\frac{7}{10}$
20. $-4.8, -4.08, -4\frac{1}{8}$ — $-4.8, -4\frac{1}{8}, -4.08$
21. $5.78, -5\frac{7}{8}, -5.9$ — $-5.9, -5\frac{7}{8}, 5.78$
22. $\frac{6}{7}, \frac{5}{6}, 0.8$ — $0.8, \frac{5}{6}, \frac{6}{7}$
23. $2\frac{2}{3}, 2.65, 2.69$ — $2.65, 2\frac{2}{3}, 2.69$
24. $-0.38, -\frac{3}{8}, -0.04$ — $-0.38, -\frac{3}{8}, -0.04$
25. $\frac{3}{7}, 0.45, \frac{4}{9}$ — $\frac{3}{7}, \frac{4}{9}, 0.45$

26. Twenty-four karat gold is considered pure. If a ring is 14 karat gold, what is its purity as a fraction and as a decimal rounded to the nearest thousandth?

    $\frac{7}{12}$; 0.583

27. Amy scored 9.02 on her balance beam routine. Joella scored 9.2 on her balance beam routine. Which gymnast received the higher score?

    Joella

LESSON **2-11**

Practice B

Comparing and Ordering Rational Numbers

Compare the fractions. Write < or >. Justify your answer.

1. $-\frac{7}{8}$ [<] $-\frac{5}{8}$ — $-7 < -5$
2. $\frac{3}{10}$ [<] $\frac{3}{8}$ — Tenths are smaller than eighths.
3. $4\frac{7}{12}$ [>] $5\frac{5}{12}$ — $\frac{7}{12} > \frac{5}{12}$

Compare the decimals. Write < or >. Justify your answer.

4. −0.531 [<] −0.513 — −3 hundredths are less than −1 hundredth.
5. 0.73 [>] 0.073 — 7 tenths are greater than 0 tenths.
6. $3.\overline{59}$ [<] 3.599 — $3.\overline{59} = 3.5959\ldots$; 5 thousandths are less than 9 thousandths.

Order the numbers from least to greatest.

7. $\frac{4}{9}, 0.4, 0.45$ — $0.4, \frac{4}{9}, 0.45$
8. $1.7, 1.65, 1\frac{2}{3}$ — $1.65, 1\frac{2}{3}, 1.7$
9. $3.18, 3\frac{1}{8}, 3.80$ — $3\frac{1}{8}, 3.18, 3.80$
10. $-5, -5.25, -5\frac{2}{5}$ — $-5\frac{2}{5}, -5.25, -5$
11. $-6\frac{3}{4}, 6.34, -6.4$ — $-6\frac{3}{4}, -6.4, 6.34$
12. $\frac{11}{12}, \frac{8}{9}, 0.91$ — $\frac{8}{9}, 0.91, \frac{11}{12}$
13. $-\frac{3}{5}, -\frac{5}{7}, -0.65$ — $-\frac{5}{7}, -0.65, -\frac{3}{5}$
14. $0.3, 0.345, \frac{1}{3}$ — $0.3, \frac{1}{3}, 0.345$
15. $-0.75, \frac{7}{8}, -\frac{5}{8}$ — $-0.75, -\frac{5}{8}, \frac{7}{8}$

16. A ream of paper contains 500 sheets of paper. Norm has 373 sheets of paper left from a ream. Express the portion of a ream Norm has as a fraction and as a decimal.

    $\frac{373}{500}$; 0.746

17. The density of Venus, compared to Earth having a density of 1, is 0.943. The density of Mercury is 0.983, compared to the density of Earth. Which planet has a greater density, Venus or Mercury?

    Mercury

LESSON 2-11
Practice C
Comparing and Ordering Rational Numbers

Compare the fractions. Write < or >. Justify your answer.

1. $-4\frac{3}{4}$ [>] $-4\frac{5}{6}$ $\quad -\frac{9}{12} > -\frac{10}{12}$, so $-\frac{3}{4} > -\frac{5}{6}$

2. $\frac{9}{10}$ [<] $\frac{10}{11}$ $\quad \frac{99}{110} < \frac{100}{110}$

3. $2\frac{2}{5}$ [>] $2\frac{3}{8}$ $\quad \frac{16}{40} > \frac{15}{40}$ so $\frac{2}{5} > \frac{3}{8}$

Compare the decimals. Write < or >. Justify your answer.

4. 1.031 [<] 1.3 — 0 tenths are less than 3 tenths.

5. −4.25 [>] −4.5 — −2 tenths are greater than −5 tenths.

6. $2.\overline{17}$ [>] 2.17 — $2.\overline{17} = 2.1717...$ and $2.17 = 2.170$; 7 thousands are greater than 0 thousands

Order the numbers from least to greatest.

7. $\frac{9}{16}$, 0.54, 0.65 — 0.54, $\frac{9}{16}$, 0.65

8. 3.49, 3.14, $3\frac{4}{9}$ — 3.14, $3\frac{4}{9}$, 3.49

9. −3.05, −3.5, $-3\frac{1}{10}$ — −3.5, $-3\frac{1}{10}$, −3.05

10. $-2\frac{7}{12}$, 2.59, −2.7 — −2.7, $-2\frac{7}{12}$, 2.59

Choose the greater number.

11. $\frac{4}{5}$ or $\frac{5}{6}$ — $\frac{5}{6}$

12. 0.095 or −0.9 — 0.095

13. $-\frac{1}{4}$ or −0.025 — −0.025

14. −5 or −5.01 — −5

15. A horsepower is the power needed to lift 33,000 pounds 1 foot in 1 minute. It is equivalent to 746 watts. Express 525 watts as a portion of a horsepower both as a fraction and a decimal rounded to the nearest thousandth.
$\frac{525}{746}$, 0.704

16. The density of Jupiter, compared to Earth having a density of 1, is 0.24. The density of Saturn is 0.124, compared to the density of Earth. Which planet has a greater density, Jupiter or Saturn?
Jupiter

LESSON 2-11
Reteach
Comparing and Ordering Rational Numbers

You can use fraction strips to compare two fractions that have different denominators.

Compare $\frac{2}{3}$ and $\frac{1}{2}$.

$\frac{2}{3}$ is greater than $\frac{1}{2}$.

$\frac{2}{3} > \frac{1}{2}$

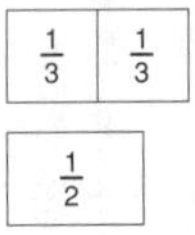

Compare $\frac{2}{5}$ and $\frac{3}{4}$.

$\frac{2}{5}$ is less than $\frac{3}{4}$.

$\frac{2}{5} < \frac{3}{4}$

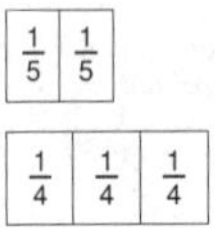

Use fraction strips to compare. Write < or >.

1. $\frac{1}{3}$ [<] $\frac{1}{2}$

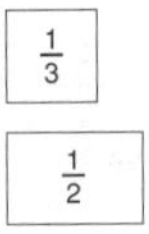

2. $\frac{3}{4}$ [>] $\frac{5}{8}$

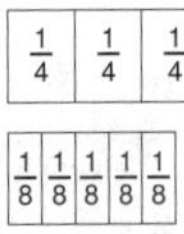

3. $\frac{1}{2}$ [<] $\frac{3}{5}$

4. $\frac{1}{5}$ [>] $\frac{1}{8}$

5. $\frac{3}{8}$ [>] $\frac{1}{3}$

To compare decimals:

1) Add zeros so that each decimal has the same number of decimal places.

2) Ignore the decimal points and compare the numbers as if they were integers.

Compare 4.3 and 4.27.

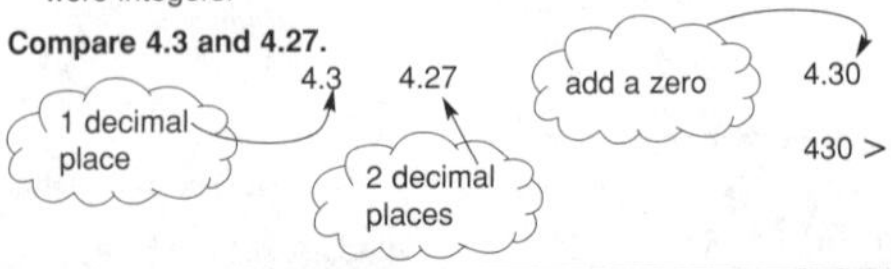

430 > 427, so 4.3 > 4.27.

Compare the decimals using < or >.

6. 7.2 [<] 7.8

7. 0.64 [>] 0.604

8. 2.09 [<] 2.817

LESSON 2-11
Challenge
Another Rational Number

Between any two rational numbers, there are other rational numbers.

Examples: $\frac{2}{3}$ lies between $\frac{1}{2}$ and $\frac{5}{6}$.
−6 lies between −5.9 and −6.1.
$1\frac{1}{4}$ lies between 1 and $1\frac{1}{2}$.

Write a rational number that lies between each pair of rational numbers below. Possible answers given.

1. 0 and $\frac{1}{5}$ — $\frac{1}{10}$

2. 3.4 and 3.5 — 3.41

3. 1.08 and 1.8 — 1.1

4. −7 and −7.5 — −7.2

5. −12.1 and −12.2 — −12.15

6. 0 and $-\frac{1}{2}$ — $-\frac{1}{4}$

7. $1\frac{1}{8}$ and $1\frac{1}{2}$ — $1\frac{1}{4}$

8. $-2\frac{1}{10}$ and $-2\frac{1}{2}$ — $-2\frac{1}{5}$

9. 3.01 and 3.02 — 3.015

10. 4.2 and $4\frac{1}{2}$ — 4.3

11. −1 and −1.1 — −1.01

12. 3.5 and $3\frac{7}{8}$ — 3.6

13. $-2\frac{3}{8}$ and −2.8 — −2.5

14. 0 and −0.3 — −0.1

15. $7\frac{1}{10}$ and 7.5 — $7\frac{1}{4}$

16. $5\frac{7}{8}$ and $5\frac{11}{12}$ — $5\frac{9}{10}$

17. $-\frac{2}{3}$ and $-\frac{3}{4}$ — $-\frac{7}{10}$

18. $3\frac{1}{5}$ and 3.19 — 3.195

LESSON 2-11
Problem Solving
Comparing and Ordering Rational Numbers

Write the correct answer.

1. Some of the wettest summers in the Northeast had the following amounts of precipitation, in inches: $14\frac{7}{10}$, $14\frac{1}{4}$, $14\frac{4}{5}$, and 14.69. Order the amounts of precipitation from least to greatest.
$14\frac{1}{4}$, 14.69, $14\frac{7}{10}$, $14\frac{4}{5}$

2. The U.S. soybean crop decreased each year from 2001 to 2003. Production for each year was approximately $2\frac{3}{4}$, 2.9, $2\frac{2}{5}$ billion bushels. Match the soybean crop, in billions of bushels, with its year.
2001: 2.9; 2002: $2\frac{3}{4}$; 2003: $2\frac{2}{5}$

3. Mark's batting average last year was 0.334. Ken's was 0.304, and Ty's was 0.343. Write the batting averages in order from lowest to highest.
0.304, 0.334, 0.343

4. Kiki lives 0.097 miles from the school, and George lives 0.131 miles from the school. Who lives farthest from the school?
George

Choose the letter for the best answer. Use the graph for problems 5–6.

NFL Field Goal Success Rate

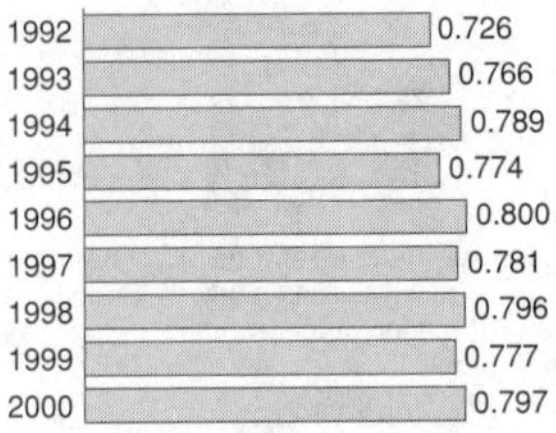

5. In which year did NFL teams have the greatest success rate?
A 1992
B 1994
(C) 1996
D 2000

6. In which year was the success rate of NFL teams greater than 0.775 and less than 0.78?
F 1994
G 1995
H 1997
(J) 1999

LESSON 2-11

Reading Strategies
Compare Visually

You can compare fractions that have the same **denominator** by comparing their **numerators.**

Which is greater, $\frac{4}{9}$ or $\frac{5}{9}$?

4 ninths   5 ninths

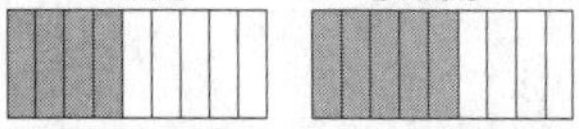

Notice that both fractions are naming ninths.
Look to see which fraction has the most ninths. ⟶ $\frac{5}{9} > \frac{4}{9}$

You can compare fractions visually by using fraction strips.

$\frac{1}{2}$	$\frac{1}{2}$

$\frac{1}{4}$	$\frac{1}{4}$	$\frac{1}{4}$	$\frac{1}{4}$

$\frac{1}{8}$	$\frac{1}{8}$	$\frac{1}{8}$	$\frac{1}{8}$	$\frac{1}{8}$	$\frac{1}{8}$	$\frac{1}{8}$	$\frac{1}{8}$

$\frac{1}{16}$	$\frac{1}{16}$	$\frac{1}{16}$	$\frac{1}{16}$	$\frac{1}{16}$	$\frac{1}{16}$	$\frac{1}{16}$	$\frac{1}{16}$	$\frac{1}{16}$	$\frac{1}{16}$	$\frac{1}{16}$	$\frac{1}{16}$	$\frac{1}{16}$	$\frac{1}{16}$	$\frac{1}{16}$	$\frac{1}{16}$

Compare the following fractions. Use the fraction strips above to help you compare. Write >, <, or =.

1. $\frac{1}{2}$ [>] $\frac{1}{8}$
2. $\frac{2}{16}$ [<] $\frac{3}{8}$
3. $\frac{3}{4}$ [<] $\frac{7}{8}$
4. $\frac{4}{16}$ [=] $\frac{2}{8}$
5. $\frac{8}{16}$ [=] $\frac{2}{4}$
6. $\frac{2}{4}$ [>] $\frac{2}{16}$

LESSON 2-11

Puzzles, Twisters & Teasers
What's the Big Deal?

Use < or > to show which number in each pair is greater. Then use the Answer Key to answer the riddle.

1. 3.8 __>__ 3.6
2. 0.088 __<__ 0.109
3. −4.26 __<__ −4.266
4. −1.902 __<__ 0.920
5. 0.755 __>__ 0.7
6. 1.82 __>__ 1.6
7. 2.25 __>__ 2.05
8. 2.88 __>__ −2.98
9. 2.05 __<__ 2.25
10. 2.88 __>__ −2.98
11. 0.32 __>__ 0.088
12. 0.999 __<__ 1.0
13. −0.93 __<__ 0.2
14. 23.44 __>__ 23
15. −0.405 __>__ −0.45

Answer Key

<	>
O S T C	I H A E

What is an insect's least favorite subject in school?

M __O__ __T__ __H__ __E__ M __A__ __T__ __I__ __C__ __S__

  < < > >   > < > < <